HYBRID INTELLIGENT ENGINEERING SYSTEMS

ADVANCES IN FUZZY SYSTEMS — APPLICATIONS AND THEORY

Vol. 1: Between Mind and Computer: Fuzzy Science and Engineering
(*Eds. P.-Z. Wang and K.-F. Loe*)

Vol. 2: Industrial Applications of Fuzzy Technology in the World
(*Eds. K. Hirota and M. Sugeno*)

Vol. 3: Comparative Approaches to Medical Reasoning
(*Eds. M. E. Cohen and D. L. Hudson*)

Vol. 4: Fuzzy Logic and Soft Computing
(*Eds. B. Bouchon-Meunier, R. R. Yager and L. A. Zadeh*)

Vol. 5: Fuzzy Sets, Fuzzy Logic, Applications
(*G. Bojadziev and M. Bojadziev*)

Vol. 6: Fuzzy Sets, Fuzzy Logic, and Fuzzy Systems: Selected Papers by Lotfi A. Zadeh
(*Eds. G. J. Klir and B. Yuan*)

Vol. 7: Genetic Algorithms and Fuzzy Logic Systems: Soft Computing Perspectives
(*Eds. E. Sanchez, T. Shibata and L. A. Zadeh*)

Vol. 8: Foundations and Applications of Possibility Theory
(*Eds. G. de Cooman, D. Ruan and E. E. Kerre*)

Vol. 10: Fuzzy Algorithms: With Applications to Image Processing and Pattern Recognition
(*Z. Chi, H. Yan and T. D. Pham*)

Vol. 11: Hybrid Intelligent Engineering Systems
(*Eds. L. C. Jain and R. K. Jain*)

Forthcoming volumes:

Vol. 9: Fuzzy Topology
(*Y. M. Liu and M. K. Luo*)

Vol. 12: Fuzzy Logic for Business, Finance, and Management
(*G. Bojadziev and M. Bojadziev*)

Vol. 13: Fuzzy and Uncertain Object-Oriented Databases: Concepts and Models
(*Ed. R. de Caluwe*)

Vol. 14: Automatic Generation of Neural Network Architecture Using Evolutionary Computing
(*Eds. E. Vonk, L. C. Jain and R. P. Johnson*)

Advances in Fuzzy Systems — Applications and Theory Vol. 11

HYBRID INTELLIGENT ENGINEERING SYSTEMS

Editors

L C Jain
University of South Australia

R K Jain
University of Adelaide

Published by

World Scientific Publishing Co. Pte. Ltd.

P O Box 128, Farrer Road, Singapore 912805

USA office: Suite 1B, 1060 Main Street, River Edge, NJ 07661

UK office: 57 Shelton Street, Covent Garden, London WC2H 9HE

Library of Congress Cataloging-in-Publication Data

Hybrid intelligent engineering systems / editors, L. C. Jain, R. K. Jain
p. cm. -- (Advances in fuzzy systems : vol. 11)
Includes bibliographical references.
ISBN 9810228899
1. Computer-aided engineering. 2. Expert systems (Computer science) 3. Neural networks (Computer science) 4. Fuzzy systems. I. Jain, L. C. II. Jain, R. K. III. Series.
TA345.H9 1997
620'.001'1363--dc21 96-51540
CIP

British Library Cataloguing-in-Publication Data

A catalogue record for this book is available from the British Library.

Part of the royalty goes to the Indrani Jain Charity Fund for Needy Students.

Printed in Singapore by Uto-Print

PREFACE

The recent advances in the theory of intelligent systems have resulted in the fusion of knowledge-based systems, artificial neural networks, fuzzy systems and evolutionary computing techniques in solving many engineering problems. These new techniques do not compete but enhance the capacity of the other. The purpose of this book is to report the integration of these intelligent techniques and their applications in engineering.

This book contains 7 chapters. The first chapter written by Jain and Vemuri presents a general introduction to the intelligent system. It also introduces paradigms such as knowledge-based systems, neural networks, fuzzy logic systems and genetic algorithms. A number of engineering problems can be solved using these paradigms. The combined use of these systems called hybrid systems is discussed.

The second chapter written by Takagi and Lee is on the integration of fuzzy systems, neural networks and genetic algorithms in the design of engineering systems. It describes the use of neural network in the automatic design of membership function of the fuzzy system. A new approach is described to design neural network by using fuzzy logic. This approach helps in identifying the problematic portion of the neural network. The use of genetic algorithms in designing fuzzy system is discussed. A fuzzy knowledge-based system is described to control genetic parameters dynamically.

Chapter 3 written by Medsker, is on the neuroexpert architecture and applications in diagnotic/classification domain. It emphasizes the integration of neural networks and expert systems, models for hybrid systems and the roles of fuzzy logic and genetic algorithms in engineering applications including diagnostic/classification domain.

Chapter 4 by Karr and Jain is on the genetic learning in fuzzy control. The technique of using genetic algorithms (GAs) to improve the performance of fuzzy logic controllers is described in this chapter. Chapter 5 by Karr and Jain, is on the cases in geno-fuzzy control. The powerful technique of using fuzzy logic control and search capabilities of GAs is applied in engineering systems.

Chapter 6 is written by H. de Garis. It describes the use of evolutionary engineering in the design of complex engineering systems. A CAM-Brain project for implementing a cellular automata based artificial brain with a billion neurons by 2001 is discussed. It is claimed that by the year 2001 it should be possible using nano-scale electronics to evolve artificial brain containing a billion neurons.

The last chapter in this book by Katayama, Kuwata and Jain is on fusion technology of neuro, fuzzy, GA and chaos theory and applications. It presents the overview of fusion technology of neural network and chaos, fusion technology of fuzzy and chaos, and fusion technology of genetic algorithm and chaos. It also list the various applications of chaos theory and describes the use of chaos in kerosene fan heater design.

This book will be useful for application engineers, scientists and researchers who wish to improve their productivity by using the state of the art techniques in engineering system design. This book will prove useful for senior undergraduate and first year graduate students in electronic, Electrical, Computer, Mechatronics and Mechanical engineering.

This book was not possible without the contribution of a number of authors listed in the book. We are grateful to them. We would like to express our sincere thanks to Mr. Berend Jan van der Zwaag for his excellent help during the preparation of the manuscript. Thanks are due to a number of reviewers for their time and efforts. We thank Professor Dr Hirota for the opportunity to edit this book.

CONTENTS

Chapter 1

An introduction to intelligent systems

Lakhmi C. JAIN
Knowledge-Based Intelligent Engineering Systems
University of South Australia
Adelaide, The Levels, SA, 5095, Australia
etlcj@levels.unisa.edu.au

Rao VEMURI
Department of Applied Science
University of California, Davis
Livermore, CA 94550, USA
vemuri1@llnl.gov

This chapter presents a general introduction to the intelligent system's techniques such as knowledge-based systems, artificial neural network, fuzzy logic systems and genetic algorithms. A number of problems can be solved using these techniques. The combined use of these techniques is also discussed.

1.1 INTRODUCTION

A machine or an industrial system is intelligent if it is able to improve its performance or maintain an acceptable level of performance in the presence of uncertainty. The main attributes of intelligence are learning or generalization, adaptation, fault tolerance and self repair and self-organization. The ability of the machine to examine and modify its behaviour in a limited sense is usually achieved by using the following techniques.

- Expert systems also called knowledge-based systems (KBS)
- Artificial neural networks (ANNs)
- Fuzzy logic systems (FLS)
- Genetic algorithms (GAs)

The combined use of these complementary techniques to solve industrial problems are becoming popular and thus paving the way for the emergence of hybrid intelligent systems. The key issue behind all these developments is the knowledge acquisition, knowledge representation and knowledge processing in intelligent environment.

The main aim of this chapter is to introduce briefly the intelligent system's techniques and their limitations and applications.

1.2 KNOWLEDGE-BASED SYSTEMS (KBS)

The knowledge-based systems are software systems which can mimic the performance of a human expert by transferring his/her expertise to the computer in a specific domain. The major logical components of an expert system are a Knowledge Base, an Inference Engine, an Interface between the system and the external environment, an Explanation facility, and a Knowledge Acquisition facility.

A "knowledge engineer" gathers the expertise about a particular domain from one or more experts, and organises that knowledge into the form required by the particular expert system tool that is to be used. Barr & Fiegenbaum list seven different forms of knowledge representation: logic, procedure, semantic network, production systems (rules), direct (analogical), semantic primitives, and frames and scripts. A typical expert system uses a data structure as the basis of the particular representation technique it implements. Consequently, the knowledge engineer needs to know the general form in which the knowledge is to be represented and the particular form of that representation required by the expert system itself. The "engineered" knowledge is the Knowledge Base.

The Inference Engine is the "driver" program. It traverses the Knowledge Base, in response to the observations and answers provided to it from the external world, in order to identify one or more possible outcomes or conclusions. The data structure selected for the specific form of knowledge representation determines the nature of the program created as the Inference Engine.

The Interface between the expert system and the external world is generally provided by a user sitting at a keyboard and screen display. However, there is increasing use of input from other sources. Sensors and transducers of many different kinds are being used to monitor the environment; these are regularly interrogated, and their output used by the expert system. Output from other computer programs, including expert systems, is often used. Extensive use is being made of vast quantities of information stored in databases to which the expert system has access. The story is similar with respect to output, which may be directed to actuators, machine controllers, other computer programs, speech synthesisers, and external data stores.

The Explanation facility is often quite simple, merely being a listing of the inputs, the appropriate parts of the Knowledge Domain, and the outcome or conclusion. Very rarely is there any facility provided to explain the reasoning process involved. In some cases, there are facilities provided to enable the knowledge engineer to provide special stores of text which may be accessed by users who need further information, either about the questions being asked or about the conclusions that are provided. These suffer from the disadvantage that they must be fully created in advance by the knowledge engineer, who must anticipate the kinds of things about which explanations might be required.

In the case of production (rule) based systems, the Knowledge Base consists of a number of rules written to an ASCII file. Therefore the Knowledge Acquisition facility required for these systems is often merely an external text editor. In systems based on other forms of representation, the Knowledge Acquisition facility will generally be an integral part of the expert system and can only be used with that system. The initial Knowledge Base is created using the appropriate facility, and modifications are made in a similar way. In very few systems does the Knowledge Acquisition facility possess any capability to "learn for itself"; the knowledge engineer is responsible for the maintenance of the Knowledge Base. The exceptions include those rule-based expert systems which have an "Induction" facility, often based on the ID3 algorithm developed by Quinlan. A series of examples is provided to the system. Each example, which is generally the result of a specific observation, consists of a set of attribute values associated with a particular outcome. Provided that enough examples are provided, a decision tree can be "induced", and this can be converted into the rule format required by the expert system being used.

Most expert systems available for use are "shells". They consist of the software programs required, but do not contain any Knowledge Bases. It is the responsibility of users to organise the creation of the required Knowledge Bases, any of which may be used by the expert system shell provided that they all satisfy the system's requirements with respect to the form of knowledge representation used and the particular structure required for the Knowledge Base. This has the advantage that the shell is usable for a large variety of applications, avoiding the necessity of creating a specific software system for each.

Advantages of knowledge-based systems

The knowledge-based systems offer advantages over the conventional programs. Some of these advantages are:

- Cost reduction in achieving a complex task
- Reduced downtime
- Capturing scarce expertise
- Flexibility in providing service
- Operation in hazardous environments
- Operation under incomplete and uncertain environment.

Limitations of knowledge-based systems

In spite of the limited success, there are difficulties associated with the knowledge-based systems. These include:

- The lack of expertise
- Expertise is hard to extract from experts
- The lack of scientific techniques in knowledge engineering

Applications of knowledge-based systems

Knowledge-based systems have been applied successfully in a large number of applications. Some of these are:

- Aerospace applications
- Design, diagnosis and monitoring of engineering systems
- Interpretation of data
- Prediction of events

Suitability of a problem for knowledge-based approach

It is very essential to choose a suitable technique for a given problem as the knowledge-based techniques are not suitable for all the problems in engineering, science and business. The suitability of a problem for a knowledge-based approach may incorporate the following ingredients.

- Task primarily requires symbolic reasoning
- Task requires use of heuristics, i.e. rules of thumb
- There exists an expert willing to work with the project
- Conventional programming approaches are not satisfactory.

Validation of the knowledge-based systems

It is not only the design of knowledge-based systems but their validation plays an important role in their ultimate usefulness. It is important to:

- Ascertain what the knowledge-based system knows or does not know or knows correctly
- Ascertain the level of expertise of the knowledge-based system
- Determine the reliability of the knowledge-based system.

1.3 ARTIFICIAL NEURAL NETWORKS (ANNs)

Artificial neural networks also called neural networks are computers containing elements that behave somewhat like the nerve cells in the brain. The study of artificial neural networks originally grew out of a desire to understand the function of the biological brain. Most of the problems are now able to be solved by means of artificial neural networks are generally also able to be solved by alternative methods. The question whether to use artificial neural networks to solve a particular problem is a matter of judgement on the part of a designer or application engineer responsible for the project. The neural network would be a suitable candidate for use if significant advantages in important areas such as cost, speed of operation, reliability, ease of maintenance, ease of initial development, ease of deployment and modification can be shown to exist.

Types of Neural Networks

A large number of neural network paradigms have been developed and used in the last four decades. Some of these widely reported and used neural network paradigms are:

- Perceptron network
- Adaline and Madaline networks
- Multiple Layer Perceptron network
- Radial Basis Function network
- Hopfield network
- Kohonen's Feature Map network
- Adaptive Resonance Theory (ART)
- Counter-propagation network
- Cognitron and Neocognitron network.

Perceptron network was introduced by Frank Rosenblatt in 1958. It is a two layer learning device using feedforward connections. It can be shown that these two layer networks are limited in applications. In 1959 Widrow introduced a basic building block called Adaline for implementing the neural networks. It consists of an adaptive linear combiner cascaded with a hard limiting quantizer. A Madaline is a two-adaline form of the structure. The multiple layer perceptron network is an extension of a single layer perceptron network. It has one or more layers usually called the hidden layers between the input layer and the output layer. The Hopfield network is a nearest match finding network. This network alters the input patterns through successive iterations until a learned pattern evolves at the output and output no longer changes on successive iterations. The Kohonen self organizing feature maps organize connections between input neurons similar to that considered to be in the brain. The neural networks based on adaptive resonance theory (ART) are capable of self-organizing the stable recognition codes in real-time in response to arbitrary sequences of input patterns. Counter-propagation networks work like look-up table in parallel for finding closest example and its equivalent mapping. The cognitron and neocognitron networks are based on the model of the brain. They are mainly designed to recognize the numerals regardless of their shape, size, distortion and position of placement.

Historical Summary

Artificial neural networks now have a relatively long history and a corresponding large amount of literature exists on their properties and development. In 1943 McCulloch and Pitts proposed neuron models based on the arrangements of hardware which attempted to mimic the performance of the single neural cell. The book written by Hebbs on 'The Organization of Behaviour' in 1949 formed the basis of 'Hebbian Learning' which forms an important part of neural network theory today. Rosenblatt constructed neuron models in hardware during 1957. These models ultimately resulted in the concept of the Perceptron. Widrow and Hoff developed first Adaline and then Madaline networks. In 1969 Minsky and Pappert's book 'Perceptron' which dampens the unrealistically high expectations previously held for neural networks. In 1982 Hopfield showed that the neural networks had potential for successful operation and applications. From this time onwards the field of neural computing began to expand and now there is a world wide enthusiasm as well as growing number of practical applications.

Why use neural networks?

The resurgence of interest in artificial neural networks over the past few years is due to a number of factors. One of the continuing and significant driving forces in neural network research has been the desire of many diverse groups is to gain an understanding of the workings and the behaviour of the brain. Some of the valuable features of artificial neural networks which distinguish this method from other algorithm based methods of computation are:

- The generalisation capability
- Parallelism in approach
- Distributed memory
- Intelligent behaviour
- Learning not programming.

Applications of neural networks

Neural networks operate in parallel and are modelled after the human brain. These networks are increasingly successfully used in various applications including areas such as business, consumer electronics, fault diagnosis, optimization, speech recognition and radar target recognition.
Anyone seeking an easy solution to a practical problem may be disappointed in their early attempts to make use of an artificial neural network. It needs to be remembered that the artificial neural network is only a special mathematical technique and like all such techniques has its limitations. Artificial neural networks appear to have their place and have proved to be very successful in certain selected applications.

1.4 FUZZY LOGIC SYSTEMS

Fuzzy logic was first developed by Zadeh in the mid 1960's to provide a mathematical basis for human reasoning. Fuzzy logic uses fuzzy set theory, in which a variable is a member of one or more sets, with a specified degree of membership. The processing of fuzzy variables involves a fuzzy logic system incorporating fuzzification, fuzzy inference and defuzzification. The fuzzification process converts a crisp value to a fuzzy input. The fuzzy inference process is responsible for drawing conclusions from the knowledge base. The defuzzification unit converts the fuzzy control actions into nonfuzzy control action.

Applications

Fuzzy logic systems have been applied successfully in a large number of applications. Some of these are automobiles, business, consumer electronics, diagnosis, engineering and industrial control.
There are some problems associated with fuzzy logic systems. For example, the conventional fuzzy logic becomes complicated if the input/output controls of a given process are complicated. Researchers are working on neural-fuzzy and

genetic-fuzzy technology which enables the finely tuned and advanced fuzzy control of process having a large number of input/output sets.

1.5 GENETIC ALGORITHMS (GAs)

Genetic algorithms were developed by John Holland in 1960's. GAs are procedures for solving problems by using principles inspired by natural population genetics. GAs use procedures to maintain a population of knowledge structures that represent candidate solutions, and then let that population evolve over time through competition (survival of the fittest) and controlled variation (recombination and mutation).

Why use genetic algorithms?

The interest in genetic algorithms and their applications grew over the past few years due to a number of reasons. Some of these are:

- GAs consider many points in the search space simultaneously and therefore have a reduced chance of converging to local minima.
- GAs work directly with the strings of characters representing the parameter set, not the parameters themselves. Thus GAs are more flexible than most search methods.
- GAs use probabilistic rules to guide search, not deterministic rules.

Applications of genetic algorithms

A large number of successful applications of genetic algorithms are reported in the literature. These include:

- Designing fuzzy logic controllers
- Job shop scheduling
- Learning the topology and weights of neural networks
- Optimisation
- Scheduling of sporting events
- Performance enhancement of neural networks
- Taste sensing and recognition.

1.6 HYBRID SYSTEMS

The intelligent systems using knowledge-based systems, neural networks, fuzzy systems and genetic algorithms have accomplished substantial gains but there are problems associated with them too. The modern trend is to integrate these techniques for offsetting the demerits of one technique by the merits of another technique. The resulting systems are called hybrid systems and the techniques used to integrate these systems is called the fusion technology. Some of these techniques are integrated as follows.

- Neural networks in designing fuzzy systems.
- Fuzzy systems for designing neural networks.
- Genetic algorithms for the design of fuzzy systems.
- Genetic algorithms in automatically training and generating neural networks.

1.7 CONCLUSION AND FUTURE DIRECTIONS

It is obvious from this introduction on intelligent systems that no single technique is capable to solve all the problems of engineering, science and business discipline. The recent research on hybrid systems suggest that by complementing each other's weakness, one can create an efficient and intelligent system for practical applications. There is no doubt in my mind that the hybrid intelligent systems will dominate the 21st century.

Bibliography

[1] Bezdeck, J.C. and Pal, S.K., Fuzzy Models for pattern recognition, IEEE Press, USA (1992)

[2] Goldberg, D.E., Genetic Algorithms, Addison-Wesley Publishing Company, USA (1989)

[3] Jain, L.C. (Editor), Electronic technology directions to the year 2000, IEEE Computer Society Press, USA (1995)

[4] Klement, E.P. and Slany, W. (Editors), Fuzzy Logic in Artificial Intelligence, Springer-Verlag (1993)

[5] Lau, C., Neural Networks: Theoretical foundations and analysis, IEEE Press, USA (1991)

[6] Liebowitz, J. (Editor), Hybrid Intelligent System Applications, Cognizant Communication Corporation, USA (1995)

[7] Mehra, P. and Benjamin, W.W., Artificial neural networks : concepts and applications, IEEE Computer Society Press, USA (1992)

[8] Morgan, N., Artificial neural networks : electronic implementation, IEEE Computer Society Press, USA (1990)

[9] Raeth, P.G., Expert Systems : a software methodology for modern applications, IEEE Computer Society Press, USA (1990)
[10] Ross, T.J., Fuzzy Logic with Engineering Applications, McGraw-Hill, Inc. USA (1995)
[11] Vemuri, V., Artificial neural networks, theoretical concepts, IEEE Computer Society Press, USA (1988)

Chapter 2

Integration of fuzzy systems and neural networks, and fuzzy systems and genetic algorithms

Hideyuki TAKAGI
Dept. of Acoustic Design
Kyushu Institute of Design
4-9-1, Shiobaru, Minami-ku, Fukuoka, 815 Japan
takagi@kyushu-id.ac.jp

Michael A. LEE
Computer Science Division
University of California
Berkeley, CA 94720, USA
leem@cs.berkeley.edu

This chapter describes four combinations of fuzzy logic, neural networks and genetic algorithms: (1) neural networks to automatic design fuzzy systems, (2) employing fuzzy rule structure to construct structured neural networks, (3) genetic algorithms to automatic design fuzzy systems, and (4) a fuzzy knowledge-based system to control genetic parameters dynamically.

2.1 INTRODUCTION

Research in the field of neural networks (NNs), fuzzy systems (FSs), and genetic algorithms (GAs) has been active since latter half of 1980s. They are categorized into a computational paradigm known as Soft Computing. According to the increasing number of Soft Computing research papers, research combining soft computing techniques has increased. In particular, NN and FS combinations have already been introduced into many consumer products [1, 18]. Each technology has different advantages. For example, the desirable properties found in soft computing techniques are: adaptability in NNs and GAs, nonlinearity in FSs and NNs, human–computer knowledge transfer in FS, implicit data knowledge dealing in NNs, global and local search properties of GAs, etc. Research combining these technologies aims to achieve high performance systems through a synergistic effect.

This chapter mainly introduces our research that combines these techniques. Section 2.2 introduces two automatic design methods for FSs using NN; section 2.3 introduces a structured NNs with an internal state can be analyzed according to the rule structure, and the problematic portion can be easily located and improved; section 2.4 introduces automatic design method for FSs using GAs, which integrates and automates three FS design stages; section 2.5 introduces a method to change GA parameters dynamically using fuzzy knowledge-based system.

2.2 NEURAL NETWORKS FOR FUZZY SYSTEMS

Although there has been much activity in techniques combining NN and FS [17], the major research direction has been to use NN for automatic designing a FS, especially membership functions, to reduce developing time and cost and increase the system performance. There are two major methods to design membership functions using an NN, which are presented in the following two subsections.

2.2.1 NN–driven Fuzzy Reasoning

NN–driven Fuzzy Reasoning [20, 21] in 1988 was the first NN approach to automatic design membership functions. This approach uses an NN to represent multi-dimensional nonlinear membership functions. The advantage of this method is that it can generate nonlinear multi-dimensional membership functions directly. In conventional FSs, one dimensional membership functions used in its antecedent parts, are independently designed, and then combined to generate multi-dimensional membership functions. It can be argued that the NN method is a more general form of the conventional FS in that the aggregated operations performed are absorbed by the NN. Conventional indirect design methods have a problem when the input variables are dependent. For example, consider an air conditioner controlled by a fuzzy controller that uses temperature and humidity as inputs. In conventional design methods of FSs, the membership functions of temperature and humidity are designed independently. The resulting fuzzy partitioning of the input space resembles Figure 2.1(a). However, when the input variables are dependent, such as temperature and humidity, fuzzy partitioning such as Figure 2.1(b) is more appropriate. It is very hard to construct such a nonlinear partitioning from one dimensional membership functions. Since NN–driven Fuzzy Reasoning constructs a nonlinear multi-dimensional membership function directly, it is possible to make the partitioning of Figure 2.1(b).

There are three design steps of NN–driven Fuzzy Reasoning: clustering the given training data, fuzzy partitioning the input space by NNs, and designing the consequent part of each partitioned space.

The first step is to cluster the training data and decide the number of rules. Prior to this step, irrelevant input variables have already been eliminated using the backward elimination [20, 21] or information criteria methods. The backward elimination method arbitrarily eliminates one of the n input variables and trains the NNs with $n-1$ input variables. The performance of NNs with n and $n-1$ is then compared. If the performance of the $n-1$ input networks is similar or better than the n input networks, then the eliminated input variable is considered irrelevant. The training data is clustered and the distribution of the data is obtained. This clustering gives rough partitioning of input space, and the number of clusters is the number of rules.

The second step is to decide the cluster boundaries from the cluster information obtained in step 1; the input space is partitioned and a multi-dimensional input membership function is decided. Supervised data is provided by the membership grade of input data to the cluster that is obtained in step 1. First an NN with n inputs and c outputs, where n is the number of relevant input variables and c is the number of clusters determined in step 1, is prepared. Training data for this network, NN_{mem} in Figure 2.2 is generated by from the clustering information given by step 1. Generally, each input vector is assigned to one of the clusters. The cluster assignment is combined with the input vector to form a training pattern. For example, in the case of four clusters and an input vector which belongs to cluster 2, the supervised portion of the training pattern will be (0, 1, 0, 0). In some cases, the user may intervene and manually construct the supervised portion if s/he believes an input data point should be classified differently than given by the clustering. For example, if the user believes that a data point belongs equally to class one and two, an appropriate supervised output pattern might be (0.5, 0.5, 0, 0). After training this NN_{mem} on this training data, the NN_{mem} computes the degrees to which a given input vector belongs to each cluster. Therefore, we assume that this NN_{mem} acquires the characteristics of the membership functions for all rules by learning and can generate the membership values that corresponds to any arbitrary input vector. The NN–driven Fuzzy Reasoning is the FS that uses a NN_{mem} as a membership generator.

The third step is the design of the consequent parts. Since we know which cluster to assign an input data to, we can train the consequent parts using the input data and the desired output. An NN expression can be used here as in [20, 21], but any other of the proposed methods, such as math equations or fuzzy variables, can be used instead. The essential point of this model is the NNs which partition the input space in fuzzy clusters.

Figure 2.3 shows one example of an NN–driven Fuzzy Reasoning system. This example is a model which outputs a singleton value computed by an NN or a TSK model [24]. Multiplication and addition in

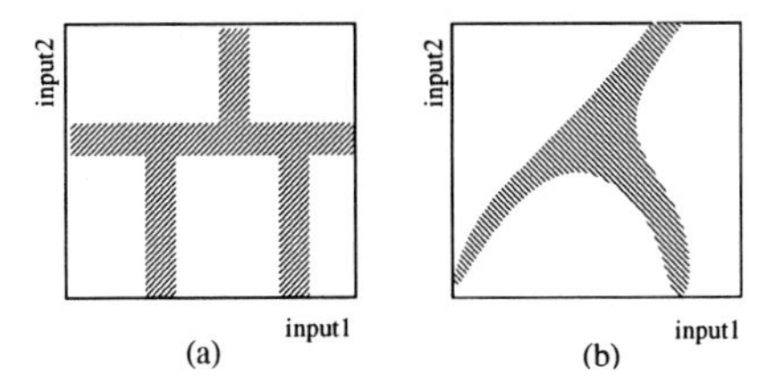

Figure 2.1: Example of fuzzy partitioning: (a) conventional (b) desired

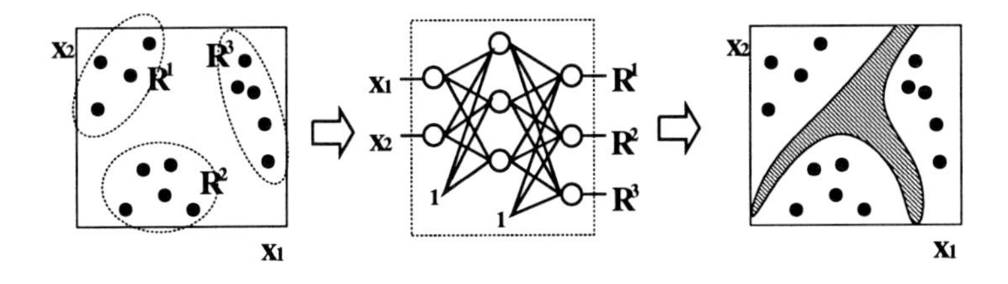

Figure 2.2: How to fuzzy partition nonlinearly by NN_{mem}

the figure calculate a weighted average. If the consequent part outputs fuzzy values, proper t–conorm and/or defuzzification operations should be used.

2.2.2 Tuning of parameterized fuzzy systems

Another NN approach to design membership function was proposed in 1989 [6]. The parameters which define the shape of the membership functions are modified to reduce error between output of the FS and supervised data. Two methods which have been used to modify these parameters are: gradient-based methods and GAs. The GA methods will be described in the next section, and the gradient based methods will be explained in this section.

The procedure of the gradient based methods are: (1) decide how to parameterize the shape of the membership functions and (2) tune the parameters to minimize the actual output of a FS and the desired

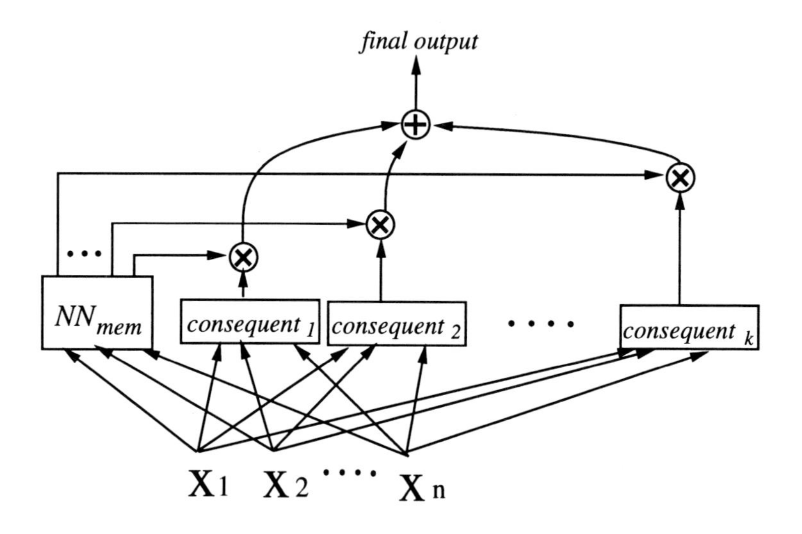

Figure 2.3: Example structure of NN–driven Fuzzy Reasoning

output using gradient methods, commonly steepest descent method. Center position and width of the membership functions are commonly used as shape definition parameters. Ichihashi *et al.* [6] and Nomura *et al.* [15, 16], Horikawa *et al.* [4, 5], Ichihashi *et al.*citeIchihashi2 and Wang *et al.* [26], and Jang [8, 9] have used triangular, combination of sigmoidal, Gaussian, and bell shaped membership functions respectively. They tune the membership function parameters using steepest descent method.

Figure 2.4 shows this method and is isomorphic to Figure 2.5. μ_{ij} in the figure is the membership function of input parameter x_j in the i-th rule, and actually it is represented by a parameter vector that describes the shape of the membership function. Namely, the method casts the fuzzy system as an NN by representing membership functions as weights and rules by nodes which perform t–norm operations. Any network learning algorithm, such a backpropagation, can be used to train this structure. y_i in the figure is a trainable output value of each consequent part. This method has already been applied in the design

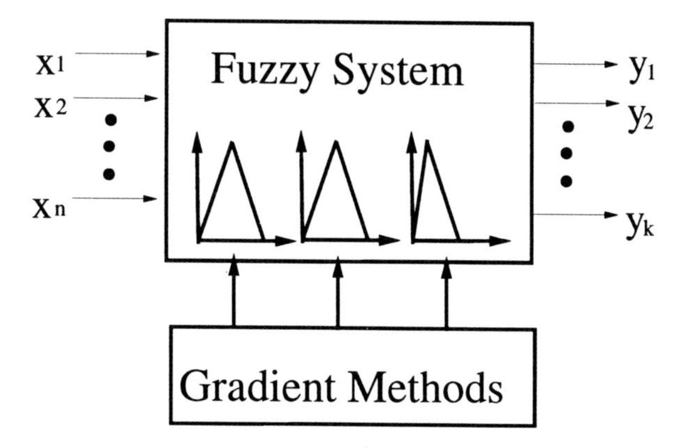

Figure 2.4: Neural networks tune the parameters of fuzzy systems

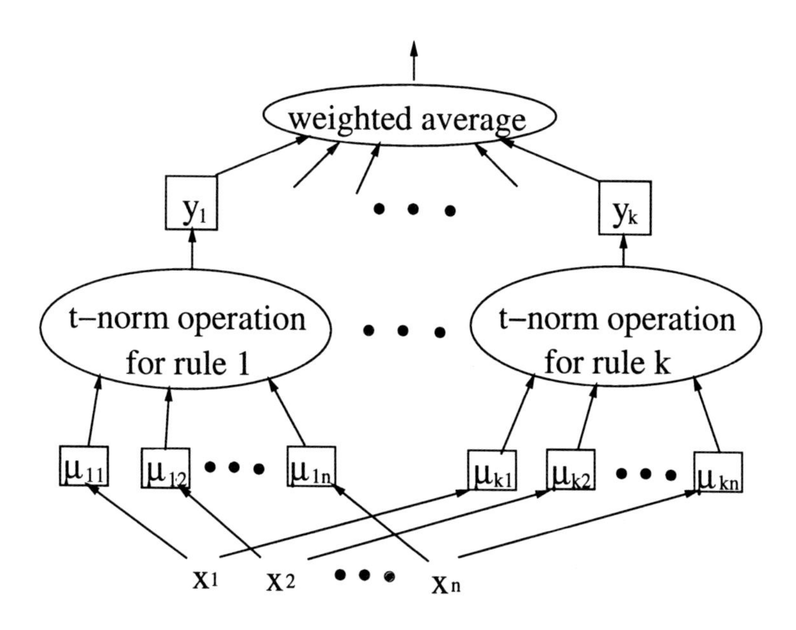

Figure 2.5: Neural networks for tuning fuzzy systems: μ_{ij} is the membership function of input parameter x_j in i-th rule

of actual commercial products [18, 1].

2.2.3 Applications

The idea of NN–driven Fuzzy Reasoning mentioned in section 2.2.1 is to use NN to realize multi-dimensional membership functions. The outputs of the NN are rule strengths which represent the combination of antecedent membership values.

This model was used in the control of the Hitachi rolling mill since 1991 [14]. The purpose of the rolling mill is to flatten plates of iron, stainless, or aluminum using 20 rolls.

The surface shape of the plate reel is detected by scanning (see Figure 2.6). The scanned shape is then fed into an NN. The NN identifies the surface pattern and outputs the similarity to standard template patterns. Since fuzzy control rules are made for each standard surface pattern, the outputs of the NN correspond to what degree an input surface pattern matches each fuzzy rule; this is, the output corresponds to rule strength. In other words, the NN takes the role of antecedent parts of all fuzzy rules. Using the aggregated final output of the fuzzy system, the 20 rolls are controlled to make the plate flat at the scanned line.

The second approach described in section 2.2.2 has been applied to develop many commercial products. The first neuro–fuzzy consumer products were made using this approach and were put on the market in 1991. Some of these products are washing machines, vacuum cleaners, rice cookers, clothes driers, dish washers, electric vacuum pots, inductive heating cookers, oven toaster, kerosene fan heaters, refrigerators, electric fan, range–hoods, and photo copiers.

Figure 2.7 shows a fuzzy system that was used in copy machines by Matsushita Electric [19]. The NN is used in the developing phase and only the final tuned fuzzy system is implemented in the products.

Consumer products based on other types of neuro–fuzzy combination are described in the reference [18].

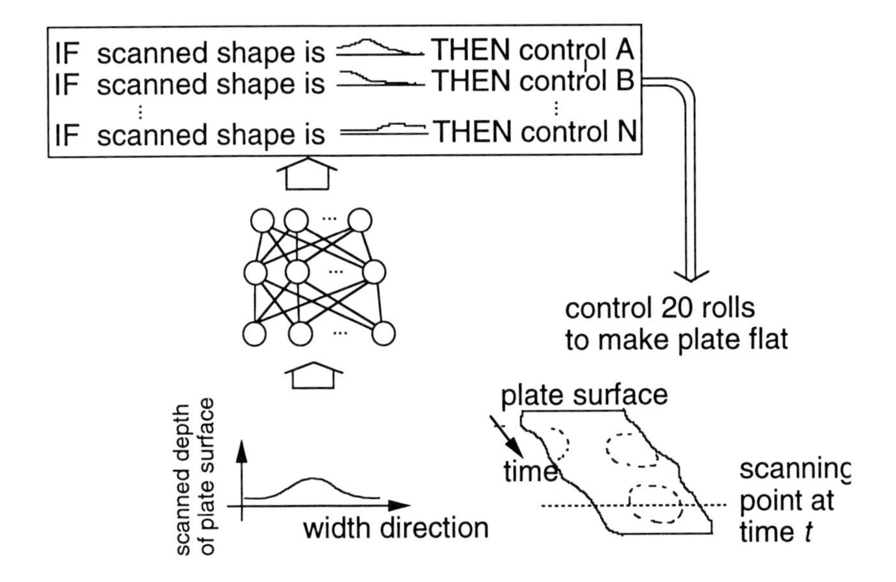

Figure 2.6: Rolling mill control by fuzzy and neural systems. The scanned pattern of plate surface is recognized by an NN. The output value of the each output unit of the NN is used as a rule strength for the corresponding fuzzy rule.

2.3 FUZZY SYSTEMS FOR NEURAL NETWORKS

This section proposes NARA (Neural Networks designed on Approximate Reasoning Architecture) [22, 23]. NNs have learning capability and can acquire knowledge from data. This is the biggest advantage of NNs. The acquired knowledge, however, is implicitly expressed in a distributed manner by synapse weights. Therefore, it is difficult to identify troublesome areas inside the NNs and improve the performance.

Performance improvement of an NN depends to the extent to which all previous knowledge can be incorporated into the NN model. NARA is a structured NNs, which incorporates knowledge structure of higher order into NNs by employing fuzzy inference rule structure. As a result,

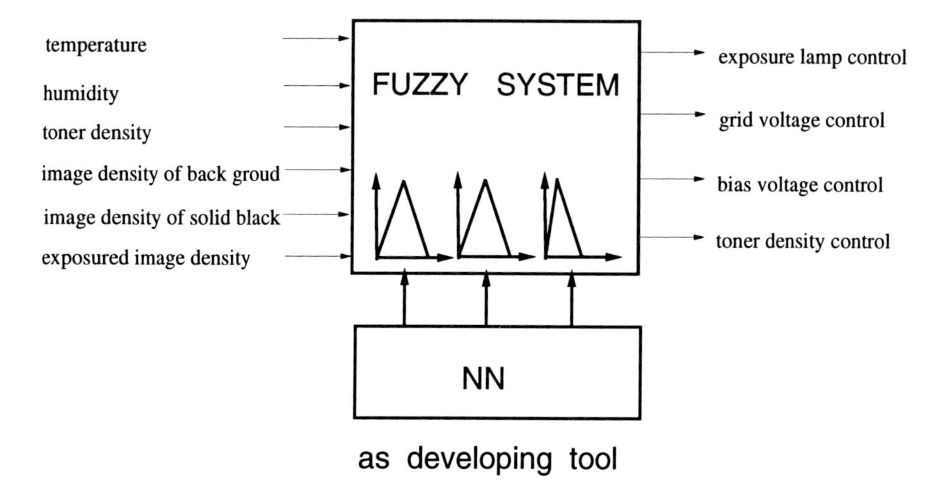

Figure 2.7: An NN designs the membership functions of copy machines.

it becomes easier to analyze, identify, and correct problems in the NNs.

2.3.1 Knowledge structure for neural networks

Suppose a feed–forward NN is applied to the classification task shown in Figure 2.8 (a). The network has two–inputs and one–output. Although the network size is very simple, it is easy to imagine that the task is so complex that it is hard for the NN to solve. However, we, human, easily find knowledge from this task; it is roughly separated into four classes like as Figure 2.8 (b). Generally, we can get this structure information by using clustering techniques.

NARA is a structured NN which is constructed based on such explicit knowledge. The strategy of a rule-based system is to partition input space and design action strategy for each partitioned space. When the system is a FS, the input space is fuzzy–partitioned. We can use the strategy to construct structured NNs; one NN roughly partitions a input space, and each NN is assigned to simple sub-task in each partitioned space. NARA shows a structure similarity to the special case of

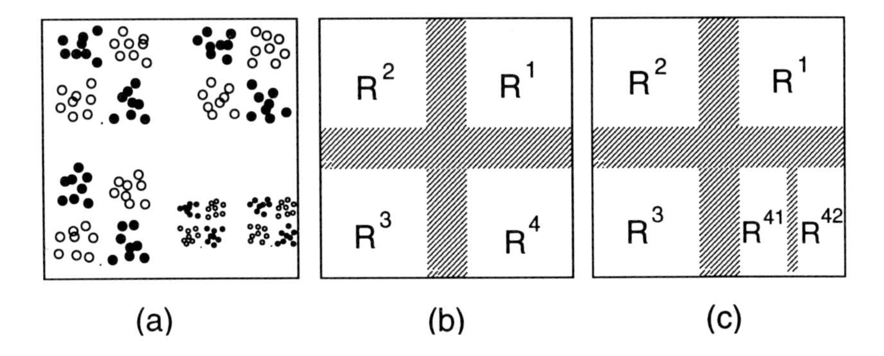

Figure 2.8: (a) A complex classification task sample: (b) and (c) fuzzy partitioning of (a)

NN–driven Fuzzy Reasoning which is introduced in section 2.2.1. In the case of Figure 2.8(b), The NARA is constructed according to four IF–THEN rules. One NN classifies four rule areas of R^1, R^2, R^3, and R^4; four NNs which correspond to consequent parts identify pattern in each area where each NN has responsibility. The point of NN–driven Fuzzy Reasoning is to incorporate an NN to design membership functions of fuzzy inference rules automatically and generate membership values in a FS. On the other hand, the point of NARA is to allow easier analysis of a NN for the purpose of performance improvement by incorporating knowledge structure into the NN structure.

2.3.2 Internal analysis to improve capability

Let's see how we can analyze NARA and improve the performance. When the performance of the NARA does not reach to the user's anticipated and required accuracy, the error will be analyzed to improve the performance. In the case of Figure 2.8 (b), the NARA has one NN which roughly partitions an input space and four NNs which identify local patterns. As easily we can imagine, the performance of the NN that identifies the pattern at the area of R^4 is poorer than other four NNs.

Table 2.1: Comparison of identification capability [23]

	conventional NN	pre-improved NARA	post-improved NARA
training data	50%	85%	94%
test data	50%	83%	89%

Then analysis is conducted by clustering the pattern date in R^4 as the area is divided into two subclasses (see Figure 2.8 (c)). Now we can introduce a new NN that partitions R^4 into R^{41} and R^{42}, and two new NNs which become active when input data is in R^4. There will possibly be various cases other than the method to cope with the issues by production–rule base style.

Experimentally the performance was actually increase by introducing knowledge structure into NNs [23]. 200 training data sets are used to train NARA and evaluated with 500 test data sets. Single conventional NN which has 57 synapse weights could not identify; NARA that has same number of weights has 35% and 33% higher identification rate than the conventional NN. By analyzing and improving the NARA, we get 9% and 6% improvement.

2.3.3 Applications

NARA has been used for a FAX ordering system. When retail electric shop orders goods from a Matsushita Electric dealer, they write an order form by hand and send it by FAX. The FAX machine at the dealer site passes the FAX image to a NARA. The NARA recognizes the hand-written characters and sends character codes to the delivery center (Figure 2.9).

This FAX ordering system must have high recognition rates, because it is used for business customers. This is main reason why NARA is used, because of its high character recognition performance; NARA became one of three winners at the public character recognition con-

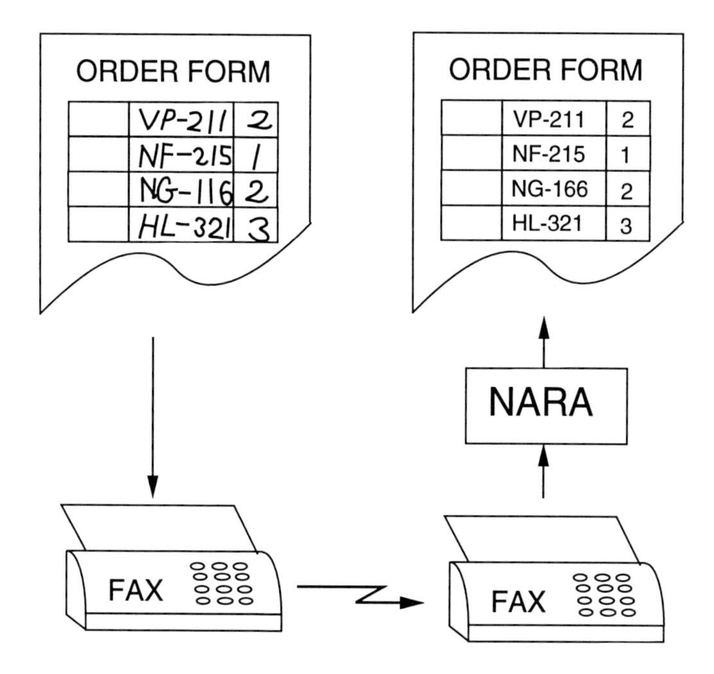

Figure 2.9: FAX OCR: hand-written character recognition system for a FAX ordering system.

test sponsored by National laboratories under the Japanese Ministry of Posts and Telecommunications [25]. Currently, the FAX–OCR part is on the market.

2.4 GENETIC ALGORITHMS FOR FUZZY SYSTEMS

GA approaches to designing FSs first appeared in 1989 [10]. In this section, we introduce a GA approach that addresses three FS design decisions simultaneously: determining membership function shape, number of rules, and consequent parameters [11]. This section will first give a short discussion of GA, followed by how to apply GA to FS design, and finally some results based on this technique.

GAs are population-based search methods that employ mechanisms inspired by natural genetics. Solutions, or parameter sets, are genetically coded as symbolic strings of 'genes'. New solutions are generated from existing solutions by operating on the genetic material contained in the population. Three common operations in a GA are: selection, crossover, and mutation. The main mechanism that drives the search process is survival of the fittest. In GA, the assumption is that any information responsible for an individual's ability to thrive in an environment is contained in its genetic code. Therefore, the quality of the genetic material is directly reflected in its ability to perform in the environment. Higher fit individuals are selected to reproduce more often than less fit individuals. The manner in which the selection operation is carried out determines the balance of exploration vs. exploitation. Crossover is a mixing operator that combines genetic material from selected parent solutions. Mutation is usually an unary operator, i.e. one that takes information from a single individual to produce another individual. An example mutation operator for binary genetic codes is one that randomly changes the genetic code from 1's to 0's and vice versa.

2.4.1 GA coding for designing FSs

Using GA involves five main steps: determine the solution space, formulate a genetic coding of points in the solution space, design genetic operators, determine the fitness function to drive the selection, and determine the genetic operator parameters (i.e., crossover rate and mutation rate). In this section, we concentrate mainly on the system representation and fitness function formulation. As a demonstration of a GA for designing FS, we use a TSK type FS system with triangular membership functions, MIN t–norm, and weighted sum t–conorms. A TSK model FS is one whose consequent part of rule is a linear combination of the antecedent values, i.e.,

$$\text{IF } \theta \text{ is } \textit{small} \text{ and } \frac{\delta\theta}{\delta t} \text{ is } \textit{big} \quad \text{THEN } \textit{force } w_0 + w_1\theta + w_2\frac{\delta\theta}{\delta t}$$

As an example FS representation that parameterizes the shape of membership functions, number of rules, and consequent parts, we propose the following:

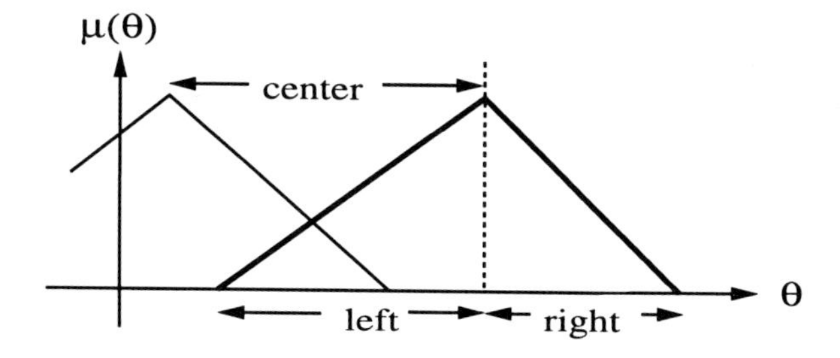

Figure 2.10: Membership function representation

center	left base	right base
10100110	10011000	01011000

membership function chromosome (MFC)

w_0	w_1	w_2
10100110	10011000	01011000

rule–consequent parameters chromosome (RPC)

Figure 2.11: Composite chromosomes

1. represent membership functions by left base width, right base width, and center distance from the previous membership function's center,

2. represent the number of rules indirectly by considering the boundary conditions of the application and the center positions of membership functions (that is, rules that involve membership functions whose center values are outside of the application boundary values are ignored), and

3. represent consequent parts as coefficients in the linear equation (see Figures 2.11 and 2.10).

A straightforward genetic coding is to map these parameters directly into a concatenated binary encoding (see Figure 2.12).

The issue of deciding the fitness function used to determine the goodness of a solution is highly application dependent. As an example,

fuzzy variable θ			fuzzy variable $\delta\theta/\delta t$			rule–consequent parameters		
MFC_1	$\cdots$	MFC_{10}	MFC_1	$\cdots$	MFC_{10}	RPC_1	$\cdots$	RCP_{100}

Figure 2.12: Gene map

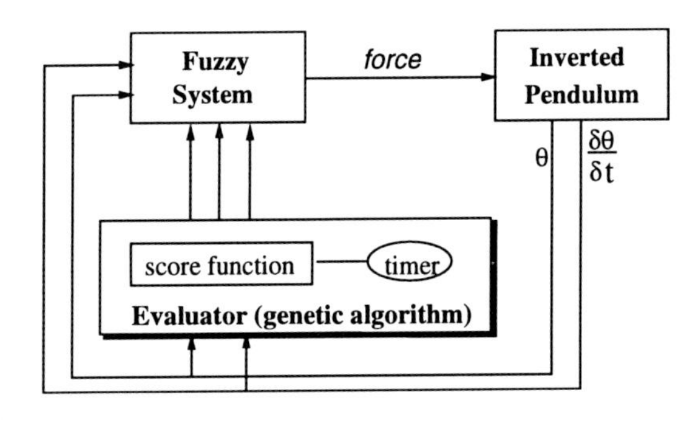

Figure 2.13: Auto-design of fuzzy controller for pole pendulum

we will design a fuzzy controller for an inverted pendulum. The inverted pendulum is a classical control task in which the objective is to balance a pole on a cart placed on a fixed length track. A continuously variable force can be applied either to the left or right side of the cart (see Figure 2.13). A trial begins by initializing the pole and cart to a particular state and then letting the system continue until (1) the pole falls over or the cart reaches the end of the track (bad), (2) the simulation time runs out (good), or (3) the pole becomes balanced (best). The fitness of a FS is determined by evaluating it on several trials, which use different initial conditions spread over the input space (this is an attempt to ensure the robustness of the final controller). Figures 2.14 and 2.15 show the performance of a FS designed with this method [11].

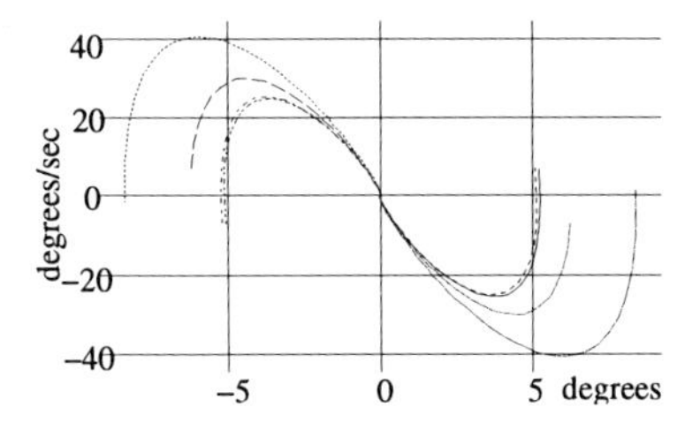

Figure 2.14: θ and $\delta\theta/\delta t$ trajectory

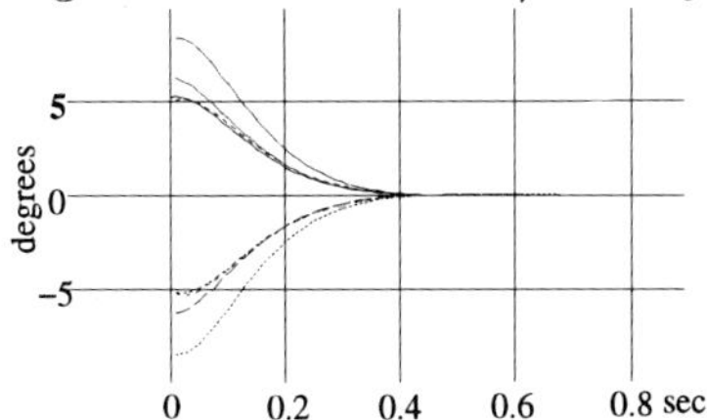

Figure 2.15: θ displacement vs. time

2.4.2 Embedding *a priori* knowledge

One of the key strengths of FS is the ease with which a designer can transfer knowledge gained from experience into a FS representation: embedding *a priori* knowledge. In this section, we show how our GA-based FS design technique can take advantage of this property[12]. There are two techniques for embedding *a priori* knowledge into the design method; via genetic and system representations or via population initialization. System representation can incorporate application knowledge such as symmetry or FS knowledge such as constraining the amount of membership function overlap. Population initialization uses known solutions to seed the initial population rather than starting with random solutions. For example, evenly distributing the membership functions over the input space can work well in situations where the system response is linear with respect to the inputs. Another technique is to initialize the consequent parts according to a known control law.

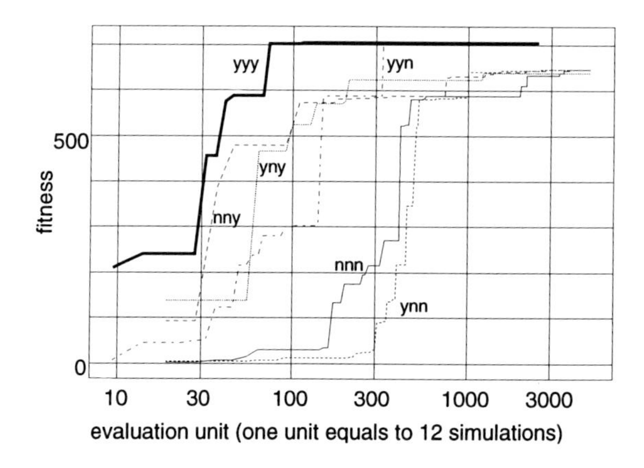

Figure 2.16: Evaluation of embedding *a priori* knowledge: the detail of the legend is given in Table 2.2.

Table 2.2: Combination of *a priori* knowledge

	nnn	nny	ynn	yny	yyn	yyy
symmetric membership			○	○	○	○
symmetric consequent					○	○
two heuristic initializations		○		○		○

Figure 2.16 shows the fitness level as a function of time for different combinations of *a priori* knowledge.

2.4.3 Remarks

In this section, we have demonstrated GA-based methods for FS design. The technique outlined is able to determine membership function shapes, number of rules, and consequent parameters in a single design phase. Because the GA search method does not require gradient information, structure and parameter space can be explored simultaneously. The genetic representation presented uses a fixed length description, which introduces constraints on the upper limit on the number of rules.

While it did not pose a problem in the example presented, in more complex systems this may prove to be disadvantageous. As pointed out earlier, the ability to embed *a priori* knowledge is a key point in using FS as knowledge structures. Although the inclusion of structural knowledge provides the most advantage, it can be detrimental if the knowledge is incorrect. What is most interesting is that the inclusion of knowledge through intelligent initialization can lead to satisfactory solutions without over-constraining the solution space.

2.5 FUZZY SYSTEMS FOR GENETIC ALGORITHMS

This section introduces the Dynamic Parametric GA: a GA that uses a fuzzy knowledge base to dynamically control GA parameters (see Figure 2.17). As noted in the previous section, one of the steps in applying GA to a problem is to set the genetic operator parameters. For example, the user must specify population size, crossover rate, and mutation rates. However, the setting of these parameters has historically been ad hoc as the complex relationship between GA performance and GA parameter settings is not well understood. In the Dynamic Parametric GA approach, we consider the GA as a dynamic process. The objective of this process is to output good solutions. Therefore if we characterize the behavior of the search process by its ability to find good solutions, we can use a meta-level search technique to optimize the parameters of the search process with respect to the ability to obtain good solutions. To further extend this notion, we contend that the best search performance will be achieved if the parameters are allowed to change dynamically (traditional GAs will not adapt the parameters during a run). We propose using fuzzy techniques for controlling the dynamic search process. The choice representing control strategies using FSs was made not only because it is easy to capture expert knowledge in them, but because of the available automatic design and optimization techniques, such as the method mentioned in the previous section. In addition, the knowledge contained in the FS is potentially recoverable, by humans, after applying design techniques. This feature makes it

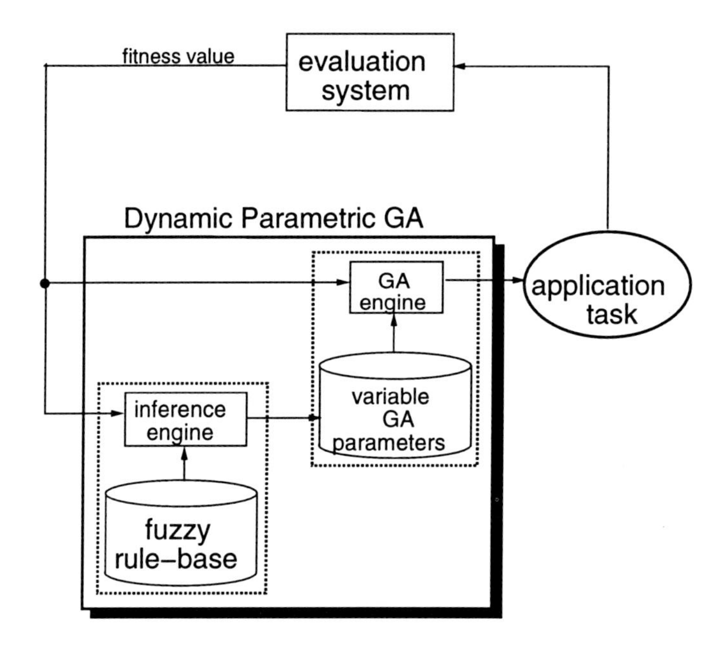

Figure 2.17: Dynamic Parametric GA.

possible to extract new knowledge about the relationship between GA parameters and behavior. The following sections introduce the framework of the Dynamic Parametric GA, discusses and presents techniques for Designing the fuzzy knowledge–base used in the Dynamic Parametric GA, and then evaluates the performance of it on an application task.

2.5.1 Dynamic Parametric GA

This section introduces the framework for the Dynamic Parametric GA. A FS is used to output GA parameter settings given the state of the GA search process. Example inputs could be (average fitness)/(best fitness), current population size, or current mutation rate. Rules in the fuzzy knowledge base would be able to reason about these inputs and provide a control action. For example, the following rules may appear in the fuzzy knowledge–base:

IF (average fitness)/(best fitness) is *big*	THEN increase population size.
IF (worst fitness)/(average fitness) is *small*	THEN decrease population size.
IF mutation is small AND population is *small*	THEN increase population size.

Rules may be used to dynamically control the exploration/exploitation trade-off. For example, when the average fitness approaches the best fitness, the population can be seen as converging. In this case, increasing the mutation rate can lead to greater population diversity.

2.5.2 Designing the Dynamic Parametric GA

In the previous subsection, we proposed the framework to control GA parameters dynamically. The next question is how to design the fuzzy knowledge-based system, which plays an essential role in the Dynamic Parametric GA. While it may be possible to manually design the fuzzy knowledge-based system for GA control, it can be rather difficult because of the complex interaction between control parameters. By employing the fuzzy technique presented in the previous section, we can attempt to design an high performance fuzzy knowledge-based system for the Dynamic Parametric GA. Figure 2.18 shows the diagram. The final part of this section presents performance results of the Dynamic Parametric GA design using this automatic technique.

As in the cart–pole application presented in the previous section, to apply the GA-based FS design technique presented in section 2.4, we must first specify a FS representation and specify how to evaluate the Dynamic Parametric GA. The system representation used in these examples is very similar to the one used in the inverted pendulum example except it uses conventional rules rather than TSK rules and adjacent membership functions are constrained to fully overlap. The consequent part is coded such that a rule can selects exactly one or none of the output fuzzy sets for a given output dimension. In our system, we used the measures as inputs and outputs shown in Figure 2.19.

To evaluate the performance measure, we use search performance measures developed in similar work [3]. The search measures are called online and offline search measures. The online measure outputs the

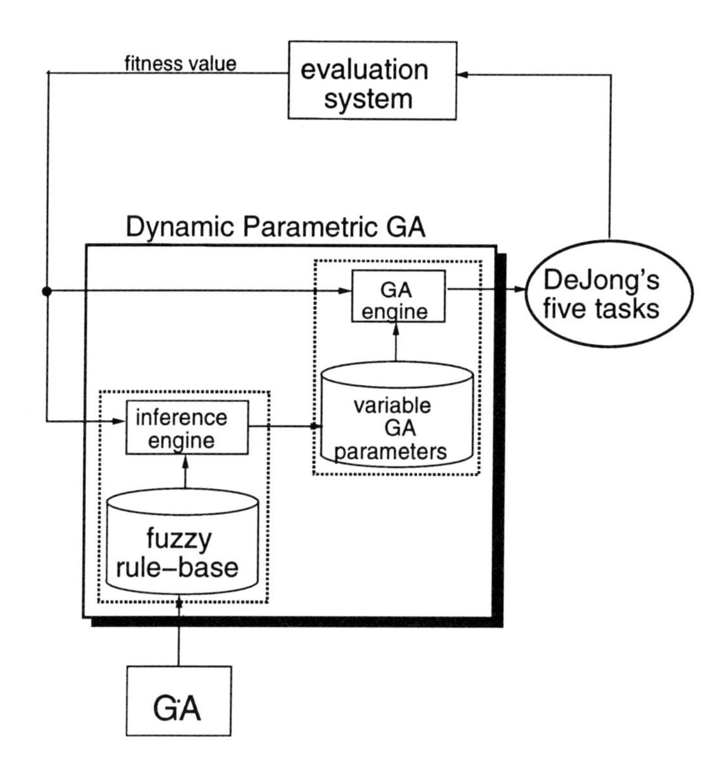

Figure 2.18: GA for automatic designing fuzzy rules in Dynamic Parametric GA.

running average of all solutions generated up to a given time. It can also be viewed as the expected value of a solution produced after a specified amount of time. The offline measure is the running average of the best solution generated up to a given point in time. Likewise, it can be viewed as the expected value of the best solution if the search were terminated at a given point in time. The overall performance measures used in our examples were based on the Dynamic Parametric GA's online and offline performance [2]. For each measure, we generated two separate Dynamic Parametric GAs, one for online and one for offline. For each measure, the performance was based on the set of DeJong functions [2]. The

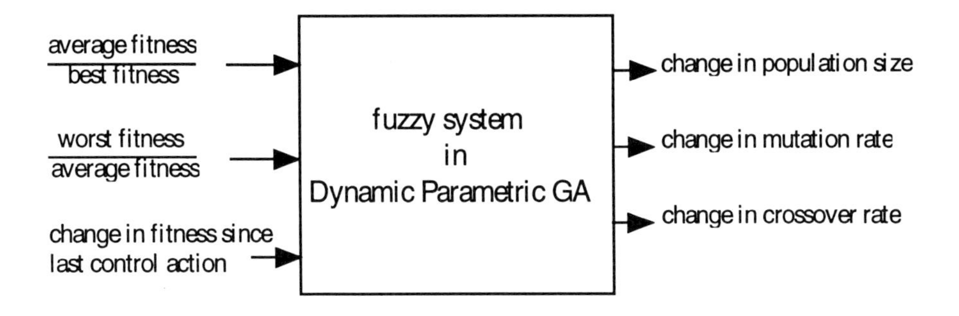

Figure 2.19: Inputs and outputs of the fuzzy system in Dynamic Parametric GA.

performance of an algorithm was actually the sum of the performance measures of the algorithm on each of the DeJong functions. The results of the meta-level optimization, i.e., the fuzzy rules and membership functions which dynamically change the population size, crossover rate, and mutation rate, are given in the reference [13].

2.5.3 Evaluation of the Dynamic Parametric GA

To validate the Dynamic Parametric GA obtained, we compared the online and offline performance of the simple static GA with Dynamic Parametric GA on the inverted pendulum task presented in the previous section. Figure 2.20 shows the experimental conditions compared. The FSs optimized for online and offline performance over the DeJong function suite in section 2.5.2 were used directly in the Dynamic Parametric GA for the inverted pendulum evaluation. Figure 2.21 shows the online performance [2] of the two GAs. The results show an improvement in GA online performance and may indicate the universal applicability of this control strategy. We hasten to point out that the fuzzy knowledge-based system in section 2.5.2 is not designed for the application task of DeJong's five functions but as a generic GA (see Figure 2.18). This is why the FS designed at section 2.5.2 works well for a different application task, fuzzy controller design for an inverted pendulum.

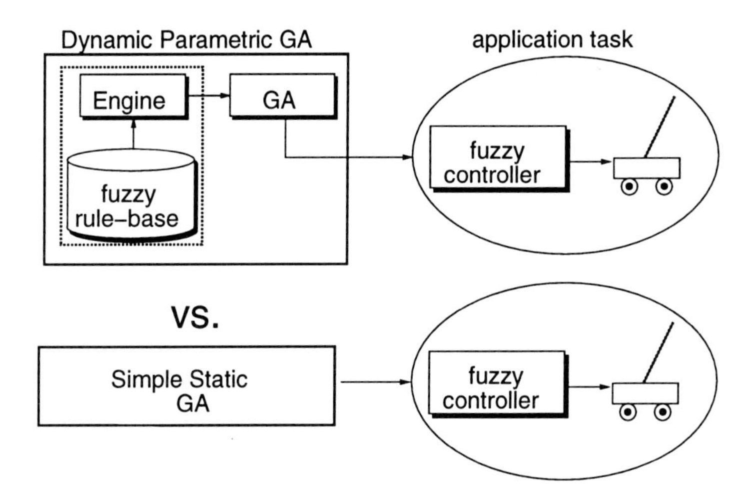

Figure 2.20: Evaluation of the Dynamic Parametric GA and a simple static GA

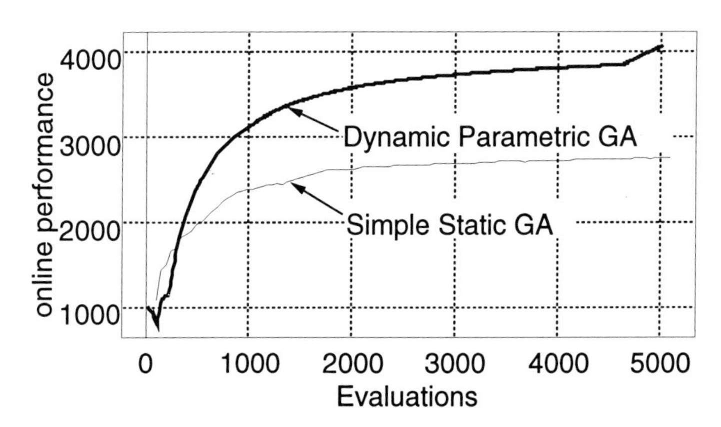

Figure 2.21: Comparison of online performance of the Dynamic Parametric GA.

2.5.4 Remarks

In this section we presented the framework of the Dynamic Parametric GA and techniques to design it. The system uses a fuzzy knowledge–base to represent parameter control strategies. These control strategies can be obtained and optimized using meta-level search techniques and lead to high performance search behavior. A key advantage of representing the control strategy using a fuzzy knowledge–base is that knowledge gained from learning has the potential for being recovered. The DPGA framework not only offers high performance search behavior, but also offers an avenue toward gaining insight on the complex relationship between GA control parameters and GA search behavior.

2.6 CONCLUSION

This chapter presented an overview of techniques for integrating NN, FS, and GA: NN for FS, FS for NN, GA for FS, and FS for GA. In the real world there is no perfect technology; each has its advantages and disadvantages. What we have shown is that through hybridization, we can obtain techniques that combine the advantages of more than one technology to compensate for deficiencies in the individual technologies.

Bibliography

[1] Asakawa, K. and Takagi, H., "Neural Networks Applications in Japan," *Communications of ACM*, Vol.37, No.3, pp.106–112 (1994).

[2] DeJong, K. A., "An analysis of the behavior of a class of genetic adaptive systems," Doctoral Dissertation, University of Michigan, University Microfilms 76-9381 (1975).

[3] Grefenstette, J. J., "Optimization of control parameters for genetic algorithms," *IEEE Trans. on Systems, Man, and Cybernetics*, Vol.16, No.1, pp.122–128 (1986).

[4] Horikawa, S., Furuhashi, T., Okuma, S., and Uchikawa, Y., "Composition Methods of Fuzzy Neural Networks," Int'l Conf. on Ind., Elect., Control, Instr., and Automation (IECON'90), pp.1253–1258 (Nov., 1990).

[5] Horikawa, S., Furuhashi, T., and Uchikawa, Y. "On Fuzzy Modeling Using Fuzzy Neural Networks with the Back–Propagation Algorithm," *IEEE Trans. Neural Networks*, Vol.3, No.5, pp.801–806 (1992).

[6] Ichihashi, H. and Watanabe, T., "Learning Control by Fuzzy Models Using a Simplified Fuzzy Reasoning", *J. of Japan Society for Fuzzy Theory and Systems*, Vol. 2, No.3, pp.429–437 (1990), (*in Japanese*).

[7] Ichihashi H. and Tanaka, "Backpropagation Error Learning in Hierarchical Fuzzy Models," Symposium of SICE Kansai Chapter, pp.131–136 (1990) (*in Japanese*).

[8] Jang, J–S. "Rule Extraction Using Generalized Neural Networks," 4th IFSA World Congress (IFSA'91). Vol.Artificial_Intelligent, pp.82–86 (July, 1991).

[9] Jang, J–S. "Self-Learning Fuzzy Controllers Based on Temporal Back Propagation," *IEEE Trans. Neural Networks*, Vol.3, No.5, pp.714–723 (1992).

[10] Karr, C., Freeman, L., Meredith, D., "Improved Fuzzy Process Control of Spacecraft Autonomous Rendezvous Using a Genetic Algorithm," SPIE Conf. on Intelligent Control and Adaptive Systems, pp.274–283 (Nov., 1989).

[11] Lee, M. A., and Takagi, H. "Integrating Design Stages of Fuzzy Systems using Genetic Algorithms," 2nd IEEE Int'l Conf. on Fuzzy Systems (FUZZ-IEEE'93), Vol.1, pp.612–617 (March, 1993).

[12] Lee, M. A., and Takagi, H. "Embedding Apriori Knowledge into an Integrated Fuzzy System Design Method Based on Genetic Algorithms," 5th IFSA World Congress (IFSA'93), pp.1293–1296 (July, 1993).

[13] Lee, M. A. and Takagi,H., "Dynamic Control of Genetic Algorithms using Fuzzy Logic Techniques", 5th Int'l Conf. on Genetic Algorithms (ICGA'93), pp.76–83 (July, 1993).

[14] Nakajima, M., Okada, T., Hattori, S., and Morooka, Y., "Application of pattern recognition and control technique to shape control of the rolling mill," *Hitachi Review*, Vol.75, No.2, pp.9–12 (1993) (*in Japanese*).

[15] Nomura, H. Hayashi, I. and Wakami, N., "A self-tuning method of fuzzy control by descent method," 4th IFSA World Congress, Vol. Engineering, pp.155–158 (July, 1991).

[16] Nomura, H. Hayashi, I. and Wakami, N., "A learning method of fuzzy inference rules by descent method," IEEE Int'l Conf. on Fuzzy System (FUZZ-IEEE'92), pp.203–210 (March, 1992).

[17] Takagi, H., "Fusion technology of fuzzy theory and neural networks — Survey and future directions —", Int'l Conf. on1Fuzzy Logic & Neural Networks (IIZUKA-90), Vol.1, pp.13–26 Iizuka, Japan (July, 1990).

[18] Takagi, H., "Applications of Neural Networks and Fuzzy Logic to Consumer Products," ed. by J. Yen, R. Langari, and L. Zadeh, in *Industrial Applications of Fuzzy Control and Intelligent Systems*, Ch.5, pp.93–106, IEEE Press, Piscataway, NJ, USA (1995).

[19] Takagi, H., "Survey of Fuzzy Logic Applications in Image Processing Equipment," ed. by J. Yen, R. Langari, and L. Zadeh, *Industrial Applications of Fuzzy Control and Intelligent Systems*, Ch.4, pp.69–92, IEEE Press, Piscataway, NJ, USA (1995).

[20] Takagi, H. and Hayashi, I., "Artificial_neural_network–driven fuzzy reasoning," Int'l Workshop on Fuzzy System Applications (IIZUKA'88), pp.217–218 (Aug., 1988).

[21] Takagi, H. and Hayashi, I., "NN–driven Fuzzy Reasoning," *Int'l J. of approximate Reasoning* (Special Issue of IIZUKA'88), Vol. 5, No.3, pp.191–212 (1991).

[22] Takagi,H., Kouda, T., and Kojima, Y., "Neural-network designed on approximate reasoning architecture and its application to the pattern recognition", 1st Int'l Conf. on Fuzzy Logic & Neural Networks (IIZUKA'90), Vol.2, pp.671–674 (July, 1990).

[23] Takagi, H., Suzuki, N., Koda, T., and Kojima, Y., "Neural networks designed on approximate reasoning architecture and their applications", *IEEE Trans. on Neural Networks*, Vol.3, No.5, pp.752–760 (1992).

[24] Takagi, T. and Sugeno, M., "Fuzzy Identification of Systems and Its Applications to Modeling and Control," *IEEE Trans. SMC–15–1*, pp.116–132 (1985).

[25] Toyota, A, "News flush," *J. of Inst. of Elect., Info., and Commun. Engineers*, Vol.72, No.7, p.812 (1992) (*in Japanese*).

[26] Wang, L–X. and Mendel, J.M., "Back-Propagation Fuzzy System as Nonlinear Dynamic System Identifier," IEEE Int'l Conf. on Fuzzy Systems (FUZZ-IEEE'92), pp.1409–1418 (March, 1992).

Chapter 3

Neuro-expert architecture and applications in diagnostic/classification domains

Larry R. MEDSKER
Department of Computer Science & information Systems
American University
Washington, DC 20016, USA
medsker@email.cas.american.edu

Hybrid intelligent systems are becoming a part of the repertoire of software systems developers for industrial applications of artificial intelligence. The integration of neural networks and expert systems has proven to be a useful way to develop real-world applications, including the areas of control systems, robotics, diagnostic systems, and industrial operations. This chapter emphasizes the integration of neural networks and expert systems, models for hybrid systems, and the roles of fuzzy logic, genetic algorithms, and case-based reasoning in neuro-expert systems. An emerging opportunity for integrating intelligent systems is the use as intelligent agents of distributed systems such as LAN's and the world-wide web.

3.1 INTRODUCTION

Several intelligent computing technologies are becoming useful as alternate approaches to conventional techniques or as components of integrated systems [25]. This chapter is an overview of individual intelligent technologies and how they can be integrated, with an emphasis on industrial applications of hybrid neural networks and expert systems.

Hybrid intelligent systems, as depicted in Figure 3.1, are usually implemented by means of traditional computing systems [25]. Expert systems and neural networks are well established as useful technologies that can complement each other in powerful hybrid systems [4,9,25–29,39,46]. Other technologies that have more recently been exploited in hybrid systems are fuzzy logic, genetic algorithms, and case-based reasoning [25,39,45]. Developers are finding niches for each of these, and the various combinations of the technologies are being explored and used.

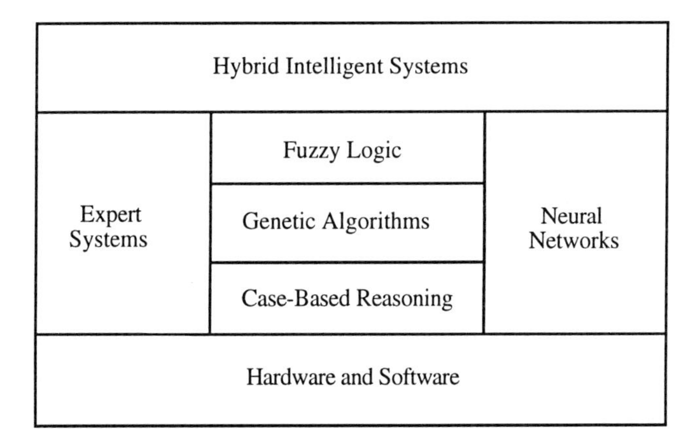

Figure 3.1: Intelligent technologies being used in hybrid intelligent systems.

Due to their fundamental natures, neural network and expert system technologies have a natural synergism that can be exploited to produce powerful computing systems. In some cases, the goal is better, more efficient and effective computing systems, making up for deficiencies in the conventional approaches. Sometimes this requires adding features associated with human intelligence such as learning and the ability to interpolate from current knowledge. The appropriate use of intelligent technologies leads to useful systems with improved performance or other characteristics that cannot be achieved through traditional methods. Developers and researchers are working to understand the appropriate use of intelligent systems, and the activity in this area is starting to produce guidelines and models for future applications.

3.2 NEURAL NETWORKS AND EXPERT SYSTEMS

Beyond their role as an alternative, artificial neural networks can be combined with expert systems to produce powerful hybrid systems. These integrated systems can also involve database and other technologies to produce the best solutions to complex problems. Expert systems and artificial neural networks have unique and to a large extent complementary features. Each approach can be equally feasible, although in some cases one may have an overall advantage over the other.

3.2.1 Expert Systems

In principle, expert systems provide a logical, symbolic approach while neural networks use numeric and associative processing to mimic models of biological systems. Expert systems [30,41] perform reasoning using previously-established rules for a well-defined and narrow domains. Rule-based systems combine knowledge bases and domain-specific facts with information from clients or users about specific instances of problems in the knowledge domains of the expert systems. Ideally, reasoning can be explained and the knowledge bases easily modified, independent of the inference engine, as new rules become known. The division between the knowledge base and the inference engine is a crucial characteristic of expert systems that enabled the creation of expert system shells that keep the knowledge base separate from the details of the reasoning mechanism. Thus, many expert systems can be developed by knowledge engineers and application-area specialists without using traditional programming. This has made the technology more widely accessible and allows developers to concentrate on capturing knowledge.

In the development process, someone in the role of a knowledge engineer works with one or more experts to formulate the knowledge base. The knowledge acquisition process, which can be difficult and time-consuming, has been identified as the bottleneck in the knowledge engineering process, and current research and development efforts are addressing the need for computer support in this area. The interface for an expert system environment gives the developer a powerful and convenient way of creating and testing the knowledge base and provides modern tools for creating an effective user interface. Another important feature of an expert system is the explanation facility, which allows that user to inquire about the reasons for particular questions being asked or about the conclusions being presented by the expert system.

Expert systems are especially good for closed-system applications for which inputs are literal and precise, leading to logical outputs. They are especially useful for interacting with the user to define a specific problem and bring in facts peculiar to the problem being solved. A limitation of the expert system approach arises from our lack of understanding about cognitive processes and the way experts actually perform the tasks they do so well. The invention of numerous techniques for representing knowledge is an important contribution of artificial intelligence, but more research is needed to understand how to mimic more closely the exact reasoning process of human experts. However, rule-based systems are popular and readily developed. For stable applications with well–defined rules, practical expert systems can provide excellent performance.

3.2.2 Neural Computing

The state of the art in neural computing [4,2,30,36] is inspired by our current understanding of biological neural networks. Today's neural computing uses a limited set of concepts from biological neural systems to implement software simulations of massively parallel processes involving processing elements interconnected in a network architecture. An important function of the artificial neuron is the production of an output response based on a weighted sum of the inputs.

Information processing with neural computers consists of analyzing patterns of activity using learned information stored as weights between node connections. A popular architecture that is available in most development shells is the multi-layered feedforward network. Three or more layers of artificial neurons are used, with one layer representing input data and one layer representing the corresponding output. Between these layers one or more intermediate, or hidden, layers contain a variable number of nodes that provide sufficient complexity to the network so that complicated, non-linear relationships between inputs and outputs can be represented. Commonly, each input node is connected to each node in the first hidden layer, and each node in a hidden layer is connected to every node in the following layer. Since a weight is associated with each connection, typical networks have a large matrix of weight values that are adjusted in the training phase so that sets of input-output pairs can be learned. Most applications use the back-error propagation algorithm, or a variation of it, for training multi-layered networks.

Multi-layered networks using back-error propagation are prime examples of supervised training. In this type of training, output vectors corresponding to specific input vectors must be supplied and the network learns the relationships. Other paradigms for supervised learning are in use, but unsupervised learning is also possible with another class of networks. In that case, the network places the input vectors into categories without desired output vectors being supplied.

A common characteristic of neural networks is the ability to classify streams of input data without the explicit knowledge of rules and to use arbitrary patterns of weights to represent the memory of categories. Together, the network of neurons can store information that can be recalled in order to interpret and classify future inputs to the network. Because knowledge is represented as numeric weights, the rules and reasoning process in neural networks are not readily explainable in terms of the particular values of the weights. However, accurate performance can be demonstrated using carefully-chosen test data sets.

Neural networks have the potential to provide some of the human characteristics of problem solving that are difficult to simulate using the logical, analytical techniques of expert system and standard software technologies. For example, neural networks can analyze large quantities of data to establish patterns and characteristics in situations where rules are not known and can in many cases make sense of incomplete or noisy data. These capabilities have thus far proven too difficult for traditional symbolic or logic-based approaches.

3.2.3 Hybrid Systems

As described in [29], the integration of neural and expert systems can be viewed according to five models (see Figure 3.2). The standalone approach uses each technology separately to study an application, develop a good design, and validate the design regardless of the final delivery system. The transformational model uses one of the two technologies in the design process with the intention of using the other for final implementation. In some cases, such as expert networks, the second system is transformed to a

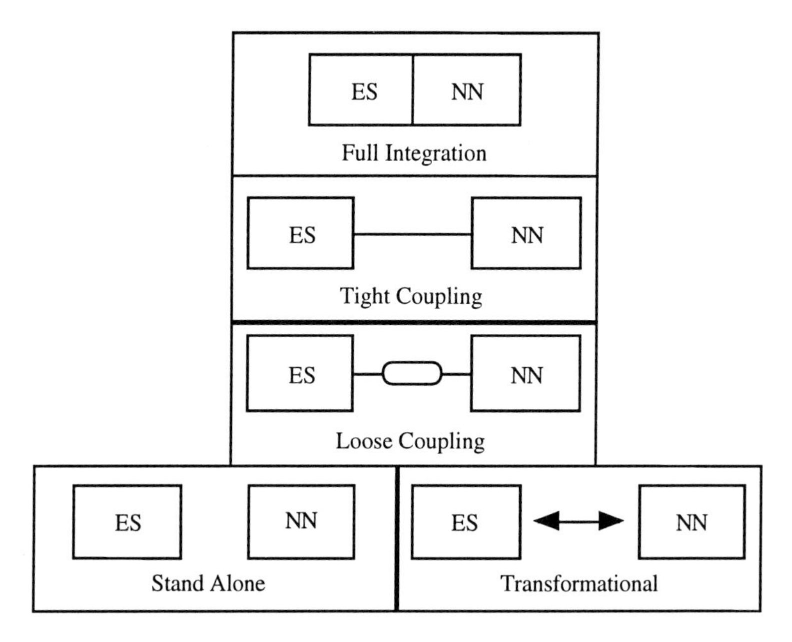

Figure 3.2: Models for integrating intelligent systems [29].

delivery system that uses the initial technology. Loose and tight coupling models have delivery systems with modules that are distinctly one intelligent technology or the other, with loose coupling achieving communication via file transfer and tight coupling using internal data structures to pass data. Loose coupling includes the use of intelligent agents in distributed systems such as LAN's and the World-Wide Web. Fully integrated systems merge the intelligent technologies, which lose their identities to produce a new type of system.

The same models can describe some of the hybrid system architectures that involve other intelligent technologies. For example, fuzzy rule based modules can be substituted or added to neuro-expert systems. Likewise, modules using genetic algorithms can be integrated in the same ways as described above.

Most of the published applications of hybrid neural network and expert systems use the loose or tight coupling models. In the simplest configuration, the output of the neural network (expert system) is input to the expert system (neural network) in a sequential control mechanism. An additional expert system component may be used for collecting input to the neural network and one may be used for analyzing the results. Also, many neural network and expert system modules can be embedded in an overall expert system or conventional software as functions to be called when needed, for example to analyze sensor data or look for trends in business data. The information gleaned by the neural network is then included with facts and rules in the larger reasoning process. In the early nineties, numerous applications have been developed using hybrid neural network and expert systems in a variety of topics areas.

An example of a connectionist expert system is Gallant's model which has been applied to diagnosis problems [9]. As shown in Figure 3.8, the nodes of the neural network represent specific facts or aspects of the knowledge domain. The input nodes represent different symptoms, and input values of +1, –1, or 0 indicate whether that symptom is present, absent, or not checked, respectively. Training data, consisting of symptoms with known diagnoses, is used to find the weights among the nodes that give the desired performance. Additional intermediate nodes allow the system to be more accurate and robust, suggesting treatments for the diagnosed diseases.

The connectionist expert system in effect represents the knowledge base by the weights of the neural network. In Gallant's model, an inference engine is used for further interpretation of the results and to direct questions from the user to minimize the amount of input while still allowing conclusions. The expert system aspect of this system also provides explanations of results. An advantage of this model is the ability to use files of training data to change the system behavior without knowing or rewriting the rules in the knowledge base.

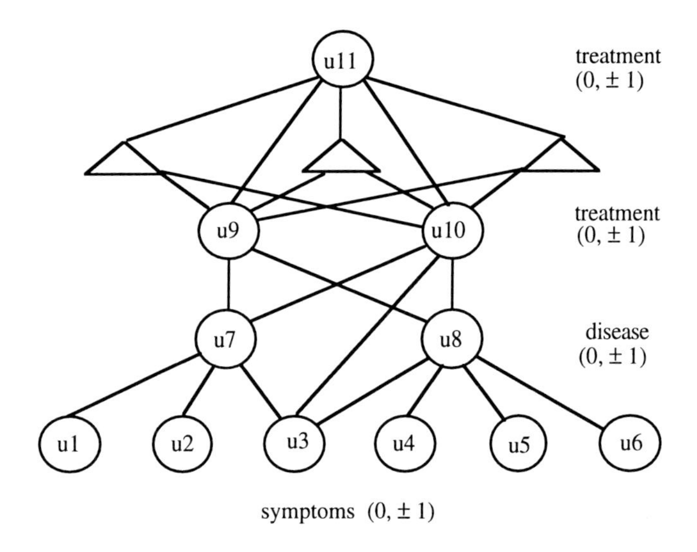

Figure 3.8: A connectionist expert system for medical diagnosis.

3.2.4 Summary

Expert systems and artificial neural networks represent complementary approaches: the logical, cognitive, and mechanical nature of expert systems and the numeric, associative, self-organizing, biological nature of neural networks. Thus, expert systems are especially good for closed systems that are literal and precise, leading to logical outputs. The value of neural network technology includes its usefulness for pattern recognition, learning classification, generalization and abstraction, and the interpretation of incomplete and noisy inputs. A natural overlap with traditional AI applications is thus in the area of pattern recognition for character, speech, and visual recognition.

Current work shows promising results for hybrid approaches in which expert systems and neural networks are used in various combinations to solve problems in a fashion more consistent with human intelligence. Interesting areas of research and development include the use of neural networks in situations where expert systems have previously been used, development of application models and guidelines for when best to use hybrid systems, and further work on creating development tools and environments.

3.3 NEURO-EXPERT SYSTEM APPLICATIONS IN INDUSTRY

Applications using hybrid systems range from general diagnostic components and the use of patterns in speech data to improve natural language processing systems to specific industrial applications for control systems and robotics. A few of the many examples of hybrid neural network and expert system applications are described below.

Table 3.1: Examples of applications of hybrid systems.

Application / Authors
Expert systems and neural networks for natural language Kwasny and Faisal [21]
A neural network-based learning system for speech processing Palakal and Zoran [33]
Workload management hybrid system Hanson and Brekke [11]
Database project planning Hillman [15]
Routing and scheduling applications Kadaba Nygard, and Juell [16]
Employee skills analysis using a hybrid system Labate and Medsker [22]

Table 3.2: Examples of specific industrial applications.

Application / Reference
Hybrid system for multiple target recognition Caglayan and Gonzalves [3]
Jet and rocket engine fault diagnosis in real time Dietz, Kiech, and Ali [7]
LAM hybrid system for window glazing design Foss [8]
SCruFFy for underwater robot welding Hendler and Dickens [13]
Chemical tank control system Hendler and Wilson [14]
Hybrid system approach to nuclear plant monitoring Mazzu, Gonsalves, and Caglayan [24]
Hybrid neural network systems for NASA ground operations Parris and Israel [34]
Hybrid systems for intelligent FMS scheduling Rabelo, Alptekin, and Kiran [35]
Aircraft flight path control Schley, Chauvin, and Mittal-Henkle [38]

In the early nineties, numerous applications had been developed using hybrid neural network and expert systems in a variety of topics areas. In Table 3.1, some examples are listed of hybrid systems in biological and medical applications.

Several industrial applications use integrated neural network and expert systems and Table 3.2 lists some examples. In some cases, the applications use the neural networks to store data for quick use in the hybrid system. For stable aspects of a problem, the neural network can be trained on tables of data for recall as needed. The expert system interface may query a user and collect client-specific information, some of which is analyzed by the neural network and some used by another expert system module or conventional software in preparing the final recommendation or report.

In the diagnosis of engine problems, patterns of fault conditions are learned by the hybrid system so that actions can be recommended in real-time. Another project uses a hybrid system architecture with a recurrent neural network to control airplane flight paths. Another application uses a hybrid system for real-time scheduling of tasks for flexible manufacturing systems. Some of the systems listed in Table 3.2 are described in more detail in the next section.

3.3.1 Working Examples

Since the beginning of the 1990's, several working hybrid neural network and expert systems have been developed and evaluated, and the following hybrid systems are examples of the many systems that have been developed. They illustrate the various models of integration and represent a variety of application topics.

Underwater Robot Welder

One of the most interesting expected uses of tightly-coupled models is in the area of blackboard architectures. Blackboards are shared data structures that facilitate interactive problem solving via independent agents. Typically the agents are knowledge-based systems, and the addition of neural networks is both technically feasible and operationally important to consider. Applications for integrated blackboard systems include complex pattern recognition, fault isolation and repair, and advanced decision support.

The SCRuFFy system by Hendler [13] uses a tight-coupling model of integrating expert systems and neural networks. The system includes a temporal pattern matcher that mediates between the two and provides a mapping from acoustic signals to symbols for reasoning about changes in signals over time. SCRuFFy uses a backpropagation neural network and an OPS5–based expert system that communicate via a blackboard architecture, which allows for future expansion to include other sensors of other types of processing modules besides expert systems and neural networks. In addition to its value as an application, SCRuFFy is a research vehicle for studying how to link symbolic and subsymbolic systems.

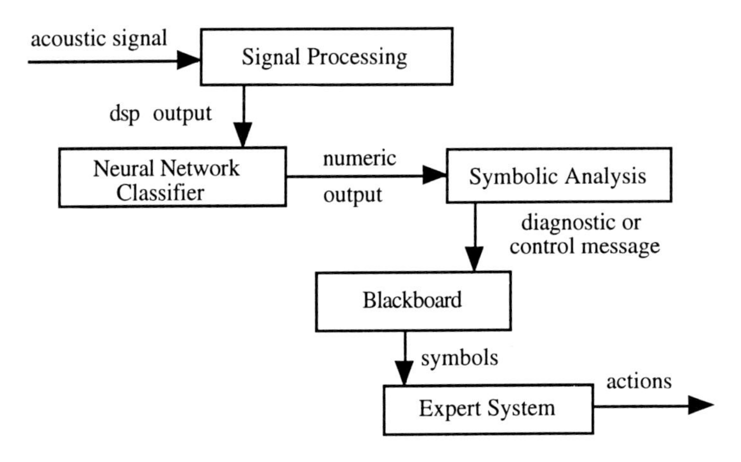

Figure 3.4: Tight-coupling hybrid system for analyzing acoustic signals.

One application of this technique is the control of the temperature of an underwater welding robot. As shown in Figure 3.4, signals from acoustic measurements from the welder are inputs to a digital signal processor that creates input to the neural network. The network is pretrained to give four numbers indicating relative classification of either normal welding or three error conditions. The symbolic analysis module tracks the changes over time in the signal classifications by the neural network and produces symbolic information describing the time course of the acoustic signal. The messages are placed on a blackboard that can be monitored by the expert system module. This information can be used by the reasoning module to recognize significant changes in the operation of the welder and recommend corrective actions early before more extreme, expensive measures are required.

LAM for Window Glazing Design

This application, called LAM, was developed at DuPont for use by architects, glazing specifiers, and laminators [8]. Laminated glass consists of two or more layers of glass, factory bonded together with an interlayer material. Uses vary from automobile safety glass to architectural glass. Design factors include safety, aesthetics, environmental insolation, and visibility. LAM has been used extensively and successfully in the field.

This integrated system of neural neural networks and expert systems was developed to facilitate the design of window glass for structural strength, hydrostatic loads, sound attenuation, and solar control. The loosely-coupled system consists of a text interface, rule-based systems, and two neural networks. The user enters design parameters for a particular window via an interactive consultation and the system critiques the design.

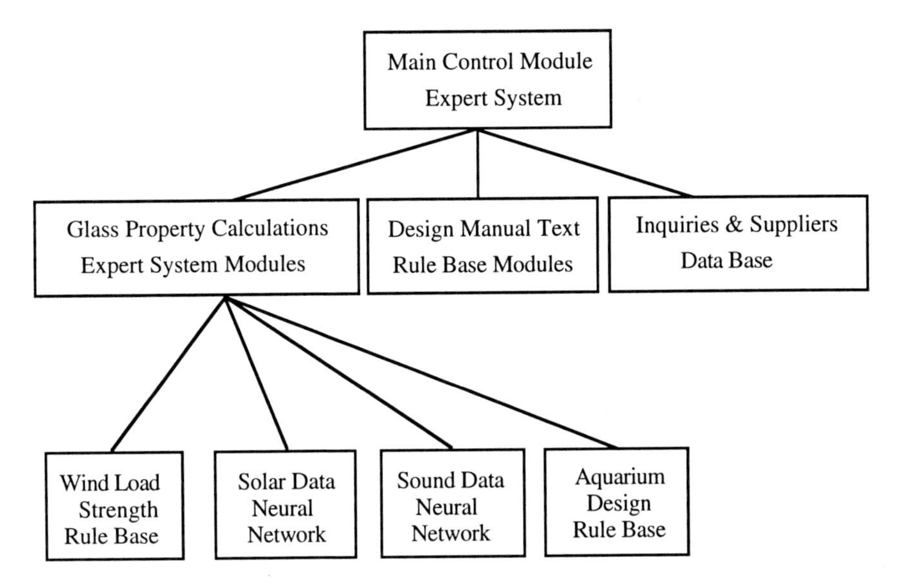

Figure 3.5: The architecture of the LAM system for window glazing design.

A diagram of the architecture of LAM is shown in Figure 3.5. The subsystems handle three different tasks: estimation of sound and solar properties, analysis of strength, and retrieval and display of text information. The text interface allows the presentation of conceptual material to the user and consists of a glass design manual and a directory of trade names and members. The logic and numerical calculations, performed by the rule-based system, address primarily the structural strength and breakage probabilities. Rule-based modules also control the overall system and process the inputs to and results from the neural networks. The neural networks are trained on published test data for solar and sound properties of different types of glass construction. The neural networks are efficient substitutes for a large number of rules or data sets and can generate estimates for values not in the test data sets.

The LAM system was developed with commercially-available software for use on personal computers. The expert system modules contain 578 rules, and the neural networks are three-layer feedforward architectures trained using the back error propagation algorithm.

Chemical Tank pH Control

The automated control of the pH value of a solution in a stirred chemical tank is a problem addressed by a hybrid neural network and expert system discussed in [14]. This application is needed, for example, in off-shore drilling in which residues must be neutralized before returning them to the ocean. Better control methods are needed for situations like this, and this hybrid system shows good potential for improvement over conventional PID controllers.

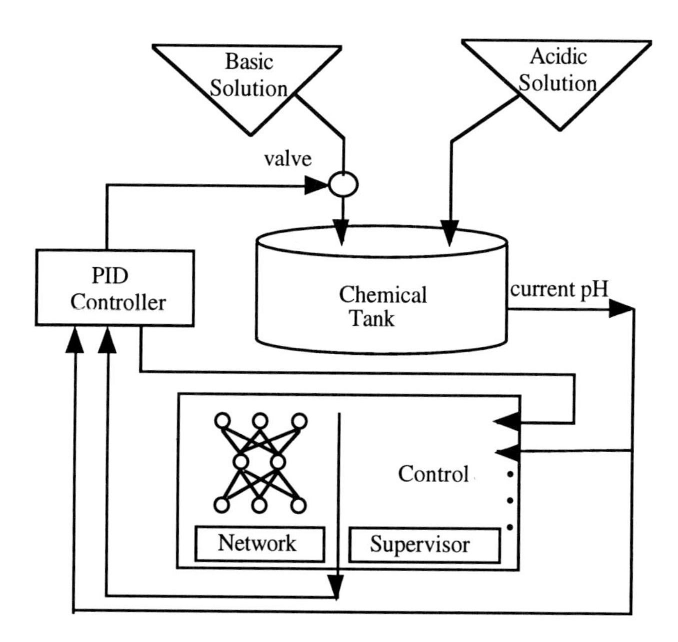

Figure 3.6: Diagram of a chemical tank pH control system using a hybrid neural network and expert system.

Shown in Figure 3.6 is their Proportional plus Derivative plus Predictive (PDP) controller, which incorporates a backpropagation neural network whose function is to predict future tank pH values. The network takes as input a history of PID controller values and tank pH values and outputs expected pH values for the tank at later time steps. A disturbance is considered to occur when the tank's pH value is out of some bound and the network predicts that it will continue to go even further. At such a time, the network's prediction is included as an additional input to the valve controller, which performs a weighted average of the PID value and the neural network predicted value to obtain better control.

To avoid over control, an expert system component decides when to include the predicted outcome in the controlling equation. The expert system examines the prediction and current behavior, and only if disturbances have occurred and the plant appears to be moving away from set point are the predictions used. The expert system uses a combination of temporal pattern matching [13] and rules to determine when to intervene.

An important aspect of this application is the use of the Conncert development software [14]. This feature allows them to introduce a high degree of automation in the application, reducing the amount of human monitoring. The software supervisor can in many cases initiate retraining of the network when better historical data needs to be incorporated.

Nuclear Power Plant Monitor

A tightly-coupled hybrid system of neural networks and expert systems was developed for a sensor monitoring system to support nuclear power plant operators [24]. Their system was created using their Macintosh-based NueX hybrid system development environment. Their systems shows good potential for enhancing operator efficiency and performance, improving plant operations by early detection of off-normal states, and improving operator training.

The architecture of their system is shown in Figure 3.7. Neural networks are employed to detect and isolate flux detector failures, which consist of subtle temporal and spatial changes. Knowledge based systems are used to determine more drastic detector failures, interpret the neural network results, and provide overall monitoring assessment.

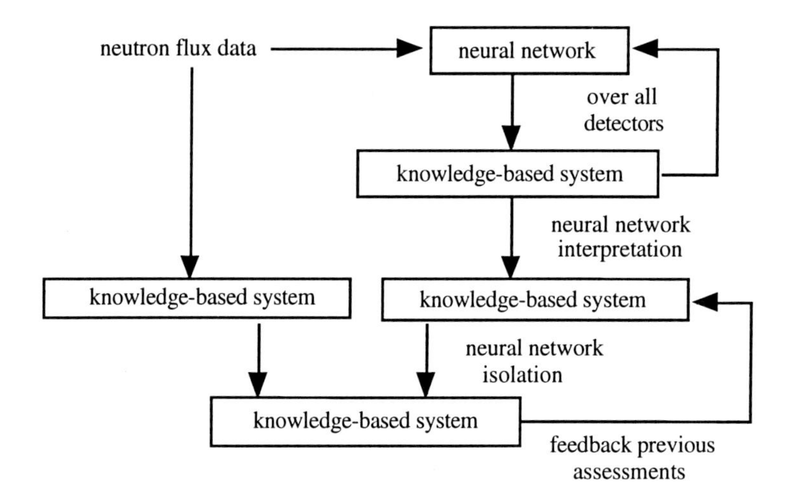

Figure 3.7: Architecture of the hybrid nuclear monitoring system.

For each weekly neutron flux data set available, the state detection neural network analyzes flux measurements and an expert system interprets the results. A database of these results is then analyzed by another expert system, taking into account detectors that are correlated. Concurrently, another knowledge-based system looks at the neutron flux data for evidence of hard failure states. The last expert system makes the final determination of each detector's operating state, presents the results through the graphical user interface, and compares the assessment with the detector's known state for system evaluation purposes.

Multiple Target Recognition

Another application developed at Charles River Analytics, Inc., performs multiple target recognition, processing and assessing multi-sensor data to find the best options to a decision maker [3]. The system learns the spatiotemporal attributes of target trajectories and classifies multi-sensor data. Their hybrid system combines conventional signal processing and probabilistic tracking algorithms with neural networks and knowledge-based modules.

As shown in Figure 3.8, at the highest level an executive expert system module performs the overall decision making, management, and coordination functions. This includes data input/output and overall target classification. The lower level consists of neural network classifiers, knowledge-based classifiers, and analytic algorithms. The target and ownship models generate simulation data for testing the system.

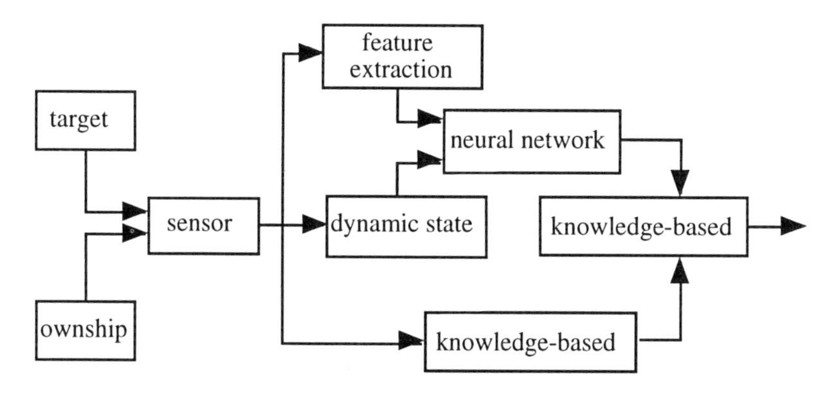

Figure 3.8: Overview of the multiple target recognition system.

Two hybrid systems were developed with different degrees of involvement of the algorithmic components and modeling. Tests on their systems show that the performance of the hybrid system is higher than that for systems based on only one of the technologies.

3.3.2 Neuro-Expert Systems Integrated with Other Intelligent Technologies

Neuro-expert systems can have components involving other intelligent technologies. Specifically, genetic algorithms, fuzzy logic, and case-based reasoning can be integrated in various combinations to produce useful industrial applications.

Fuzzy Logic in Neuro-Expert Systems

The combination of fuzzy logic and expert systems is a fundamental technique flowing directly from the nature of fuzzy logic. Fuzzy expert systems are currently the most popular use of fuzzy logic with many applications now operational in a diverse range of subjects. The fundamentals of the individual technologies have much in common. The principles of fuzzy sets and fuzzy logic were discovered [43,44,45] during the same time as the origins of expert systems; however, the slow acceptance of fuzzy techniques caused the development of knowledge-based techniques to proceed independently in a different time frame.

Expert system techniques can bring a number of advantages to a hybrid system. Logical operations can be implemented in a form that is relatively easy to understand at an applications level. The architecture of an expert system can effectively control the overall system. From the development standpoint, the developer interface is usually convenient and efficient, and the user interface for input of user requirements and presentation of results is relatively easy to construct. Expert systems should provide explanation components for justifying the overall results of the application. Fuzzy systems also have the advantage of storing knowledge of experts in a form, such as rules or mathematical expressions, that is easy to visualize, enter into the system, and modify. Furthermore, they incorporate membership functions and parameters that allow the system to be written at a more abstract level and tuned to achieve good performance. For fixed, well-defined knowledge, a fuzzy system is a convenient and effective way to represent the solution to a problem. However, for complicated and large systems, fuzzy systems become difficult to design, involving manual methods and trial and error. The matrix representing the relationships between concepts and actions becomes unwieldy, and the best values for the many parameters needed to describe the membership functions are difficult to ascertain. The performance of a fuzzy system can be very sensitive to the specific values of the parameters.

Fuzzy logic and expert system technologies are easily combined due to the common characteristics and others that complement each other. Both techniques are well suited for dealing with decision making and other knowledge-oriented problems. Systems using these techniques have enhanced performance in regard to development efficiency, system quality, and speed of execution. Several ways are possible for using fuzzy logic with expert systems, but fuzzy expert systems are the dominant form of integration [5,6,10,17,18].

In the last four years, over 400 articles have been published on the integration of fuzzy and neural systems, and fuzzy neural systems are the topics of numerous workshops and special sessions at international conferences. The prospects for useful applications are bright, amplifying the growth in fuzzy systems and neural networks individually.

Neural networks can be modified to incorporate fuzzy techniques and produce a neural network with improved performance [7,20,31]. One approach is to allow the fuzzy neural network to receive and process fuzzy input. Another option is to add layers on the front end of the network to fuzzify crisp input data to the fuzzy neural processing.

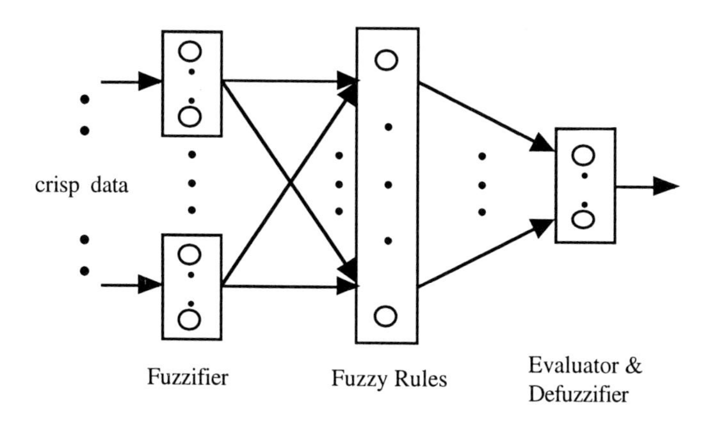

Figure 3.9: High-level view of a fuzzy neural network.

A fuzzy neural network can have three broad functions (see Figure 3.9). The initial layers process crisp input data by assigning groups of nodes to the labels of linguistic variables and implementing membership functions in nodes. Thus, crisp input data can be translated into membership function values — the output of the first layers of nodes. These values go to layers that function as fuzzy rules operating on the fuzzified input. The final layers aggregate the results of applying the rules and defuzzify the results to obtain crisp values that can become outputs of the network or that can receive further processing as part of a decision or control system. The first section of Figure 3.9 can be implemented as several layers of nodes. The nodes in the first layer can correspond to the different crisp values in the input vector and can distribute those values to sets of nodes in the second layer that represent the different linguistic variables.

Genetic Algorithms in Neuro-Expert Systems

Genetic algorithms can be used effectively with expert systems, fuzzy logic, and neural networks [1,12,19,32,37]. Hybrid systems using these combinations are the subjects of an increasing number of research and development projects.

Fuzzy and genetic systems operate well in similar environments, including situations involving nonlinearities and requiring high levels of performance. Thus, these two technologies are sometimes alternatives and can be used well in the standalone or transformational modes. They also work well as modules in coupled systems.

In hybrid systems, fuzzy components provide a clear representation of knowledge as rules or mathematical expressions. The use of membership functions and associated parameters provides flexibility and abstraction that simplifies the design of highly complex systems. Genetic algorithms facilitate the optimization of fuzzy system performance. In fuzzy system design, genetic algorithms can be used to tune membership values, prune membership functions, and derive fuzzy rules. Fuzzy logic control can be applied to the operation of genetic systems and perform the evaluation function required in genetic algorithms. A less-explored area of integration is the use of the two technologies to produce systems with more general learning abilities such as the semantic interpretation of symbols and the understanding of system behavior from their input and output data.

A natural and proven area of integration is in control systems for physical processes [19]. The improvement of controllers is an important area today, and the improvements provided by combining genetic algorithms with fuzzy systems are very attractive. The extension of hybrid genetic and fuzzy techniques to other application areas such as data analysis and control of information systems looks promising. Future work will clarify the roles of genetic algorithms and fuzzy logic in hybrid systems and provide guidelines for deciding when to use hybrid fuzzy and genetic systems.

Research and development on hybrid genetic and neural systems has grown dramatically since the late 1980's. Most of the activity has been focused on the exploitation of the advantages of genetic algorithms to improve the design and use of neural networks [12,18,42].

Most of the published work uses the ability of genetic algorithms to search large, complex spaces to prepare data for neural networks, find initial sets of parameters for training networks, and use genetic and the newer evolutionary techniques to evolve neural network topologies. Other work looks at creative ways to couple neural network modules with genetic algorithms and other problem-solving techniques.

Research and development efforts also focus on ways to represent neural networks for easier tuning and design by genetic algorithms, guidelines for coupling neural and genetic modules, ways to substitute genetic algorithms for neural learning algorithms, and opportunities for incorporating new techniques from evolutionary computing.

One way genetic algorithms can impact neuro-expert system is through the design of the neural network components. Harp and Samad [12] describe an environment called NeuroGENESYS, in which the various neural networks associated with a genetic population can be trained in parallel. They have conducted several experiments with their NeuroGENESYS system and have showed performance improvements compared to conventional manual procedures. Their method is independent of the particular type of neural network and has flexibility for the user to choose the particular optimization criteria, from network size and learning speed to suitability for hardware implementation.

As shown in Figure 3.10, populations of possible design configurations are generated and analyzed by the genetic algorithm. The chromosomes are blueprints for network design and the specific types of information they store is determined by the designer for the particular neural network model chosen. Likewise, the evaluation of the network performance can emphasize different aspects such as speed or ability to generalize. The time for evaluating the different designs is decreased by the ability to train them simultaneously.

The experimentation with this hybrid approach produced networks with, sometimes highly, improved performance. Application topics included simple digit recognition, the exclusive OR problem, process delay estimation, Kohonen feature map problems, and intelligent tutoring systems. The results are encouraging that in appropriate situations this method can effectively improve network performance.

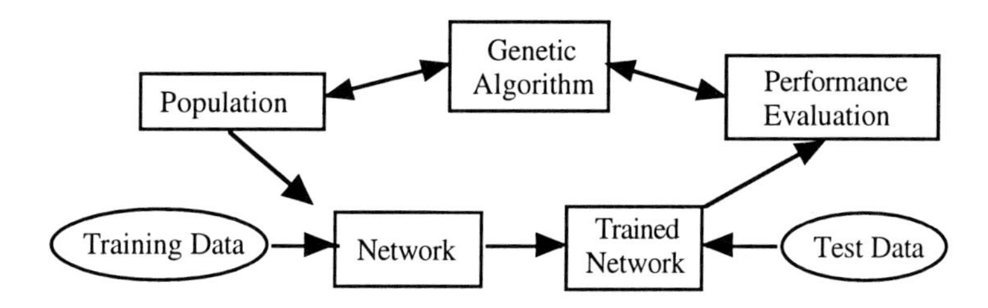

Figure 3.10: Organization of the NeuroGENESYS method for optimizing neural networks with genetic algorithms.

Relatively little work has been done on the integration of expert systems and genetic algorithms. Genetic techniques have been used to improve the performance of rule-based systems and related problem-solving techniques from the artificial intelligence and operations research fields [10,16,25].

Some work in this area uses expert systems and genetic systems to understand the knowledge domain and application details to help derive design requirements. Genetic techniques can be applied to expert networks to find better confidence factors, and they can be used to find good values of parameters in problem-solving techniques. Genetic techniques can be used to find or tune rules for an expert system, and expert systems can provide heuristics for improving the performance of a genetic system.

Neuro-Expert Systems with Case-Based Reasoning

Research and development for integrating case-based reasoning with other intelligent technologies are at an early stage. Most of the initial work has focused on combining case-based reasoning with expert system technology, which is an alternative in the same application domains. Because of the similarity in domains for these two technologies, hybrid systems involving case-based reasoning are potentially in the same areas as discussed for the combination of expert systems with fuzzy logic, neural networks, and genetic algorithms.

The other intelligent technologies can also be used at various points in the case-based reasoning process to improve performance. Knowledge-based components can be useful in the inner workings of a case-based reasoning system as part of the case retrieval and adaptation processes. Neural networks can be used to improve case-based reasoning operations in the case retrieval process, to design the initial case base, and to find a good representation technique by discovering the relevant features of cases. Genetic algorithms can be used in case-based systems operations to improve the case retrieval process and in the process of finding better indexing systems and case representation techniques. A genetic algorithm approach could potentially be used for optimizing the overall system performance.

In hybrid systems applications, the potential of integrating case-based reasoning with other intelligent technologies is to make more powerful and efficient systems by taking advantage of the strengths of each technology. The use of an expert system or a fuzzy logic component for that part of the problem that is readily amenable to description by rules can improve performance, along with providing an explanation system. The parts of a problem that involve recall of data tables or identification of patterns may be more efficiently done with neural networks. Genetic algorithms may provide a good solution to a particular part of a problem and may be part of the mechanism to optimize the performance of a case-based reasoning system.

Recent work [23] in intelligent data management in forestry uses case-based reasoning and knowledge based components in a system, Seidam, that provides answers and product information to user queries. The heart of the system is a planning system that decides the steps to follow to provide desired forest information. Sources of the information include geographical information systems, models, and image databases. Another example of a hybrid system using CBR is a shell called KwEshell for manufacturing process design [40]. The development system has been used for several industrial applications including the design of new mixing processes for a rubber company. They have used neural networks and fuzzy logic to enhance the basic CBR systems.

3.4 FUTURE FOR HYBRID SYSTEMS

Several recent developments in research and development point to the important role of hybrid neuro-expert systems in the future.

3.4.1 Soft Computing

The significance of the potential for hybrid neuro-fuzzy systems is underscored by the concept of Soft Computing [45], introduced by Lotfi Zadeh, and by the establishment of his Berkeley Initiative in Soft Computing (BISC). Rather than the traditional emphasis on precision and certainty, soft computing accommodates imprecision and uncertainty to allow reasoning and computation usually needed for practical applications. Currently, Prof. Zadeh sees soft computing as involving fuzzy logic, neural computing, and probabilistic reasoning. The latter includes genetic algorithms, belief networks, parts of learning theory, and chaotic systems. Fuzzy-neural systems are thus likely to receive considerable attention in the coming years and be important parts of practical systems.

3.4.2 Distributed Systems

Intelligent technologies can be integrated in networked environments such as LAN's and the World Wide Web. Intelligent systems can be located in various nodes of computer networks and can be integrated via the network to solve problems in a distributed fashion. Interest in intelligent agents has grown rapidly in the last few years, and all the intelligent technologies are possible as the basis for agents that work together. Communication among agents via the World Wide Web is an important development that will increase the interest in integrating intelligent technologies with each other and with conventional technologies.

3.4.3 Summary

Interest in hybrid intelligent systems continues to grow, and the initial focus on expert systems and neural networks has been greatly expanded by the developments in the individual techniques of fuzzy logic, genetic, algorithms, and case-based reasoning. The increasing numbers of useful applications and the designation of the new area called soft computing are indications of the acknowledgement and importance of hybrid systems. Whether through brain-like systems or simpler combinations of intelligent technologies, more complex systems that integrate intelligent technologies are increasingly feasible and desirable for solving practical problems. In the fields of artificial intelligence and neural networks, and from the newer areas of the other intelligent technologies, the more difficult problems are being addressed with combinations of techniques. In addition to their value for applications, hybrid intelligent systems may make significant contributions to our understanding of the brain and mind. In the area of applications, hybrid intelligent systems will find more use as development tools and environments become even more effective and economical.

BIBLIOGRAPHY

[1] Al-Attar, A., "A hybrid GA-heuristic search strategy," *AI Expert*, Vol. 9, pp. 34–37 (1994)

[2] Beale, R. and Jackson, T., *Neural Computing*, Bristol, England: Adam Hilger, (1990)

[3] Caglayan, A. K., and Gonzalves, P. G., "Hybrid system for multiple target recognition," in [Medsker, L. R., *Hybrid Neural Network and Expert Systems*. Boston: Kluwer Academic Publishers (1994)] at pp. 139–179.

[4] Caudill, M. and C. Butler, *Naturally Intelligent Systems*, Cambridge,

MA: MIT Press (1990).

[5] Cox, E., "Applications of fuzzy system models," *AI Expert* , Vol. 6, pp. 34–39 (1992)

[6] Cox, E., "Integrating fuzzy logic into neural nets," *AI Expert* , Vol. 7, pp. 43–47 (1992)

[7] Dietz, W. E., Kiech, E. L., Ali, M., "Jet and rocket engine fault diagnosis in real time," *Journal of Neural Network Computing*, Vol. 1, pp. 5–18 (1989)

[8] Foss, R. V., "LAM hybrid system for window glazing design," in [Medsker, L. R., *Hybrid Neural Network and Expert Systems.* Boston: Kluwer Academic Publishers (1994)] at pp. 49–75.

[9] Gallant, S. I., *Neural Network Learning and Expert Systems*, Cambridge, MA: MIT Press (1993)

[10] Hall, L. O. and Kandel, A., "The evolution from expert systems to fuzzy expert systems," in [Kandel, A.,*Fuzzy Expert Systems*, Boca Raton, FL: CRC Press (1992)] at pp. 3–21.

[11] Hanson, M. A., and Brekke, R. L., "Workload management expert system – combining neural networks and rule–based programming in an operational application," *Proceedings of the Instrument Society of America*, Vol. 24, pp. 1721–26 (1988)

[12] Harp, S. A. and Samad, T., Optimizing neural networks with genetic algorithms. *Proceedings of the American Power Conference*, Chicago, pp. 1138–1143 (1991)

[13] Hendler, J. and Dickens, L., "Integrating neural network and expert reasoning: an example," *Proceedings of the Eighth Conference of the Society for the Study of Artificial Intelligence and Simulation of Behaviour*, Leeds, U.K., pp. 109–116 (1991)

[14] Hendler, J. and Wilson, A., "Chemical tank control system," In [Medsker, L. R., *Hybrid Neural Network and Expert Systems.* Boston: Kluwer Academic Publishers (1994)] at pp. 109–119.

[15] Hillman, D. V., "Integrating Neural Networks and Expert Systems," *AI Expert*, Vol. 4, pp. 54–59 (1989)

[16] Kadaba, N., Nygard, K. E., and Juell, P. L., "Integration of adaptive machine learning and knowledge-based systems for routing and

scheduling applications," *Expert Systems with Applications*, Vol. 2, No. 1, pp. 15–27 (1991)

[17] Kandel, A., *Fuzzy Expert Systems*, Boca Raton, FL: CRC Press (1992).

[18] Kandel, A. and Langholz, G. (eds), *Hybrid Architectures for Intelligent Systems*, Boca Raton: CRC Press (1992)

[19] Karr, C. 1991. Applying genetics to fuzzy logic. *AI Expert* , Vol. 6, pp. 38–43 (1991)

[20] Kosko, B., *Neural Networks and Fuzzy Systems: A Dynamical Approach to Machine Intelligence*, Englewood Cliffs, NJ: Prentice Hall (1992)

[21] Kwasny, S. C., and Faisal, K. A., "Rule–based training of neural networks," *Expert Systems with Applications*, Vol. 2, no. 1, pp. 47–58 (1991)

[22] Labate, F., and Medsker, L., "Employee skills analysis using a hybrid neural network and expert system," *Proceedings of the IEEE International Conference on Developing and Managing Intelligent System Projects*, Washington, DC, March 29–31, pp. 205–211 (1993)

[23] Matwan, S. *et al.*, "Machine Learning and Planning for Data Management in Forestry," *IEEE Expert*, Vol. 10, No. 6, pp. 35–41 (1995).

[24] Mazzu, J. M., Gonsalves, P. G. and Caglayan, A. K., "Hybrid system approach to nuclear plant monitoring," in [Medsker, L. R., *Hybrid Neural Network and Expert Systems.* Boston: Kluwer Academic Publishers (1994)] at 77–108.

[25] Medsker, L. R., *Hybrid Intelligent Systems.* Boston: Kluwer Academic Publishers (1995)

[26] Medsker, L. R., *Hybrid Neural Network and Expert Systems.* Boston: Kluwer Academic Publishers (1994)

[27] Medsker, L. R. (ed.), Special Issue of *Expert Systems with Applications: An International Journal*, Vol. 2 (1991)

[28] Medsker, L. R., "Design and development of hybrid neural network and expert systems," *Proceedings of the IEEE International*

Conference on Neural Networks, Vol. III, Orlando, FL, pp. 1470–1474 (1994)

[29] Medsker, L. R. and Bailey, D. L., "Models and guidelines for integrating expert systems and neural networks," in [Kandel, A. and Langholz, G. (eds)*Hybrid Architectures for Intelligent Systems*, Boca Raton: CRC Press (1992)] at 154–171.

[30] Medsker, L. R. and Liebowitz, J., *Design and Development of Expert Systems and Neural Networks*, New York: Macmillan Publishing Company (1994)

[31] Nauck, D. and Kruse, R., "NEFCON-I: An X-Window based simulator for neural fuzzy controllers," *Proceedings of the IEEE International Conference on Neural Networks*, Vol. III, Orlando, FL, pp. 1638–1643 (1994)

[32] Oliver, J., "Finding decision rules with genetic algorithms," *AI Expert*, Vol. 9, pp. 33–39 (1994)

[33] Palakal, M. J., and Zoran, M. J., "A neural network-based learning system for speech processing," *Expert Systems with Applications*, Vol. 2, No. 1, pp. 59–71 (1991)

[34] Parris, F. R., Jr., and Israel, P., "Hybrid neural network systems for NASA ground operations," *Proceedings of the IEEE International Conference on Neural Networks*, Vol. III, IEEE World Congress on Computational Intelligence, Orlando, FL, June 28–July 2, pp. 1721–1725 (1994)

[35] Rabelo, L. C., Alptekin, S., Kiran, A. S., "Synergy of artificial neural networks and knowledge-based systems for intelligent FMS scheduling," *Proceedings of the International Joint Conference on Neural Networks*, Vol. I, San Diego, CA, June 17-21, pp. 359–366, (1990)

[36] Rumelhart, D. E., Widrow, B., and Lehr. M. A., "The basic ideas in neural networks," *Communications of the ACM* , Vol. 37, pp. 87–92 (1994)

[37] Schaffer, J. D., "Combinations of genetic algorithms with neural networks or fuzzy systems," in [Zurada, J. M., Marks II, R. J., and Robinson, C. J. (eds), *Computational Intelligence: Imitating Life*,

New York: IEEE Press (1994)] at pp. 371-382.

[38] Schley, C., Chauvin, Y., and Mittal-Henkle, V., "Integrating optimal control with rules using neural networks," *Proceedings of the International Joint Conference on NeuralNetworks*, Vol. II, Seattle, WA, pp. 759–763 (1991)

[39] Soucek, B. and the IRIS Group (eds), *Neural and Intelligent Systems Integration*, New York: John Wiley and Sons (1991)

[40] Takahashi, M. et al., "Reusing Makes It Easier: Manufacturing Process Design by CBR with Knowledge Ware," *IEEE Expert*, Vol. 10, No. 6, pp. 74–80 (1995)

[41] Turban, E., *Expert Systems and Applied Artificial Intelligence*, New York: Macmillan Publishing Company (1992)

[42] Wilke, P., "Simulation of neural networks and genetic algorithms in a distributed computing environment using NeuroGraph," *Proceedings of the World Congress on Neural Networks* , Vol. I, Portland, OR, pp. 269–272 (1993)

[43] Zadeh, L. A., "Fuzzy sets," *Information and Control* , Vol. 8, pp. 338–353 (1965)

[44] Zadeh, L. A., "Fuzzy Logic," *Computer* , Vol. 21, pp. 83–93 (1988)

[45] Zadeh, L. A., "Fuzzy logic, neural networks and soft computing," *Communications of the ACM* , Vol. 37, pp. 77–84 (1994)

[46] Zurada, J. M., Marks II, R. J., and Robinson, C. J. (eds), *Computational Intelligence: Imitating Life*, New York: IEEE Press (1994)

Chapter 4

Genetic learning in fuzzy control

Charles L. KARR
Department of Aerospace Engineering & Mechanics
University of Alabama, Box 870280
Tuscaloosa, AL 35487-0280, USA
ckarr@eng.ua.edu

Lakhmi C. JAIN
Knowledge-Based Intelligent Engineering Systems
University of South Australia
Adelaide, The Levels, SA, 5095, Australia
etlcj@levels.unisa.edu.au

Researchers at the Universities of Alabama and South Australia have developed a technique that utilizes the search capabilities of genetic algorithms to enhance the process control capabilities of fuzzy logic. Genetic algorithms are search algorithms based on the mechanics of natural genetics. Fuzzy logic is a process that affords computers the capability of manipulating abstract concepts commonly used by humans in decision making. Together, genetic algorithms and fuzzy logic possess the qualities needed in adaptive control systems. This chapter describes a process by which genetic algorithms can be used to develop efficient fuzzy controllers.

4.1 INTRODUCTION

Researchers at the Universities of Alabama and South Australia have developed a technique that utilizes the search capabilities of genetic algorithms (GAs) to enhance the process control capabilities of fuzzy logic. Both GAs and fuzzy logic are paradigms of nature: fuzzy logic is a scheme that uses linguistic terms which humans often use to describe their actions [33], and GAs are search techniques based on the mechanics of natural genetics [12]. The use of fuzzy logic in industry has increased recently due, in part, to its utility in solving practical problems. Fuzzy logic controllers (FLCs) are rule-based systems that mimic a human's practical "rule-of-thumb" approach to problem solving, and have performed well in the area of process control. GAs, on the other hand, are being used more frequently due to their ability to perform highly efficient searches in large, poorly-behaved spaces. Each of these tools has strengths that, when exercised properly, can improve the performance of the other [19, 23]. In this chapter, a technique is presented for improving the performance of FLCs using GAs.

Rule-based systems, commonly called ***expert systems***, are increasingly used in practical applications of artificial intelligence. Although conventional expert systems have performed as well as humans in several problem domains [32], their use in process control has lagged behind their use in other areas. This is because humans perform process control by using imprecise rules-of-thumb that include a degree of uncertainty. The uncertainty associated with human decision-making can be incorporated into expert systems via fuzzy set theory. In fuzzy set theory, abstract concepts can be represented with ***linguistic variables*** (fuzzy terms); terms like "very high," "fairly low," and "kind of fast." The use of these fuzzy terms provides FLCs with a degree of flexibility generally unattainable in conventional rule-based systems used for process control.

FLCs have been used successfully in a number of control problems [6, 31]. These ***fuzzy expert systems*** include rules to direct the decision process and membership functions to convert linguistic variables into the precise numeric values needed by a computer for automated process control. The rule set is gleaned from a human expert's knowledge which is generally based on his/her experience. Because linguistic terms are used in their construction, writing the rule set is often a straightforward task. However, there are situations for which writing the rule set can prove challenging. An example of this would be constructing a set of rules to control a physical system for which there is no experienced human operator. Defining the fuzzy membership functions, on the other hand, is almost always the most time-consuming aspect of FLC design. Although there has been some work in the area of rule modification [28], little has been done to ease the task of defining the membership functions. In fact, this chore is often accomplished via trial-and-error.

In general, the development of methods for choosing membership functions that optimize FLC performance has received little attention. A standard method for determining the membership functions that produce maximum FLC performance would expedite FLC development. However, locating optimal membership functions is a difficult task because the performance of an FLC often changes dramatically due to small changes in either the membership functions or the rule set [19].

There are a number of efficient search algorithms that have their origins in the field of biological evolutionary theory. These algorithms are stochastic optimization algorithms and are known as evolutionary computation. The field of evolutionary computation is comprised mainly of GAs, genetic programming, and evolutionary algorithms. GAs were developed by John Holland in the mid 1970's [12], and serve as the foundation for most work in the field. These highly efficient search algorithms have been used to solve difficult problems in a number of disciplines. GAs are the focus of the research reported in this chapter. However, the genetic programming and evolutionary algorithms deserve mentions here.

Genetic programming was developed by John Koza at Stanford University and is actually an extension of GAs. Although genetic programming is derived from GAs, it uses hierarchical genetic material whose size is not defined. The chromosomes represent tree structures and special genetic operators have been developed to work on the branches of the trees. The main advantage of genetic programming over GAs is that the size of the final solution of a given problem does not need to be known in advance.

Evolutionary programming focuses mainly on the phenotypes, not on genotypes as with GAs [12]. The genetic operators work directly on the phenotype. The evolutionary algorithms center on the behavioral link between parents and offspring while GAs focus on the genetic link. A number of researchers are working on evolutionary algorithms and time will be the real judge as to which of these three techniques eventually proves to be most effective. However, for the time being, GAs have experienced the most success.

GAs are finding increasing popularity in the field of optimization. GAs are search algorithms based on the mechanics of genetics; they use operations found in natural genetics to guide their trek through a search space [9]. Empirical investigations by Hollstien [14] and De Jong [5] have demonstrated the technique's efficiency in function optimization. De Jong's work, in particular, establishes the GA as a robust search technique — one that is efficient across a broad spectrum of problems — as compared to several traditional schemes. Subsequent applications of GAs to the search problems of pipeline engineering [9], very large scale integration microchip layout [4], structural optimization [10], job shop scheduling [3], equipment design [18], and machine learning [2, 9, 13], add considerable evidence to the claim that GAs are broadly based.

The robust nature and simple mechanics of GAs make them inviting tools not only for establishing membership functions, but also for selecting rules to be used in FLCs. GAs are also potentially beneficial in the design of adaptive FLCs which alter either their rules or membership functions "on-line" to account for changes in the physical environment. In this chapter, a GA will be used for three specific tasks required in FLC design: (1) developing a rule set, (2) selecting membership functions, and (3) altering membership functions on-line to produce an adaptive FLC. To facilitate the presentation of the technique of combining GAs and FLCs, a simple liquid level system (one that is admittedly quite easily controlled) is discussed. The attributes of the system that make it easy to control are also the ones that make it an amicable forum for discussing control strategy. Namely, it is a first order, linear system with which most people are at least casually familiar. It is important to note that the emphasis in this chapter is on the application of the method rather than the control of a particular physical system. However, GAs have been used to design FLCs for physical systems that are far more difficult to control [19, 20]. In the example considered in this chapter, the GA-designed FLC (GA-FLC) outperforms an author-developed FLC (AD-FLC).

4.2 EVOLUTIONARY COMPUTING

The field of evolutionary computing is currently composed of three techniques: (1) GA, (2) genetic programming, and (3) evolutionary computing. These techniques are effective and robust. They have associated with them mathematical theories, and thus are complicated enough to warrant entire textbooks [9, 22, 27]. The intent of this section is to introduce the very basic concepts associated with each of the techniques.

GAs are based on a Darwinian survival-of-the-fittest approach in which organisms (in this case, possible solutions to the problem at hand) are placed in a competitive environment and forced to fight for survival. Each possible solution is coded as a string of characters (a chromosome) and assigned a figure of merit (a fitness function value) that represents the quality of the solution the coded parameters represents. The result is a highly effective search algorithm that is a loose model of the evolutionary processes found in nature.

The field of GAs has become quite popular, and there is a rather large group of researchers interested both in their theory and application. Thus, the field has advanced to the point that a well-accepted nomenclature has developed. Some fundamental definitions are:

phenotype	=	the potential solution to the problem at hand
chromosome	=	the representation of the phenotype in a form that can be used by the GA
genotype	=	the set of parameters encoded in the chromosome
gene	=	the non-changeable pieces of data from which a chromosome is composed
alphabet	=	the set of values a gene can take on (in this work a binary alphabet is used; a gene is either a 0 or a 1)
population	=	the collection of chromosomes that evolves from generation to generation
generation	=	a single pass from the present population to the next
fitness	=	measure of the quality of a chromosome
evaluation	=	the translation of the genotype into the phenotype and the calculation of its associated fitness

The goal of a GA is to find the optimum individual chromosome by means of a stochastic search of the solution space. The stochastic search is completed via genetic operators; and there are now a number of such operators (see [9] for details). Section 5 of this chapter provides details of the GA employed.

Genetic programming is a technique derived from GAs developed by John Koza [22]. The basic premise of genetic programming is to develop entire computer programs that can effectively solve a given problem. In genetic programming, the solutions to the problem at hand are actually computer programs. The computer programs are represented as tree structures, and the original idea was developed using the LISP language. The members of the population are thus tree structured programs, and the genetic operators have been especially tailored to operate effectively on the branches of the tree structure.

When the potential solutions to the problem are of a hierarchical tree structure themselves, genetic programming offers a more natural chromosomal representation than standard linear-string GA does. A distinct advantage of genetic programming over GA is that the size and shape of the final solution does not have to be known in advance. The tree structured chromosomes typically vary their size and shape over the course of a generation. Research has shown that genetic programming can be successfully applied to many problems in variety of domains [22].

Evolutionary algorithms represent the third technique commonly thought to compose the field of evolutionary computing. Unlike GAs which focus on the representation of the parameters in a search (genotypes), evolutionary algorithms focus on the phenotypes. There is no need for a separation between the recombination space and an evaluation space. The genetic operators work directly on the actual structure or phenotype. The structures used in evolutionary algorithms are representations that are problem dependent and more natural for the task than the general representations used for GAs. The representation used is a vector of real values, identical to a real-valued chromosome in a GA. In evolutionary algorithms, however, the real-valued numbers are seen more as traits than genes. Evolutionary algorithms, therefore, focus more on the behavioral link between parents and offspring while GAs focus on the genetic link.

4.3 PHYSICAL SYSTEM

The problem of interest in this chapter is a rectangular tank containing liquid as shown in Figure 4.1. Figure 4.1a shows a schematic of the physical system, while Figure 4.1b shows a block diagram of the control loop. The physical system includes an inflow and an outflow, both of which are adjustable using valves. Of note is the inclusion of a gate, initially in the down position, that can be raised to expand the area of the tank by a factor of four. The gate is controlled by an external agent; it cannot be operated by the controller. Thus, the characteristic performance of the liquid level system (the rate at which the liquid level rises or falls in chamber A due to changes in the valve settings of the input and output streams) can be altered dramatically by opening and closing this gate. Since a GA is used to learn fuzzy membership functions, the liquid level system must be perturbed and controlled numerous times. Therefore, a mathematical model of the physical system is adjusted, not an actual tank containing liquid. The physical system is governed by the continuity equation:

$$h^{t+1} = h^{t} + \left(\frac{Q_i - Q_o}{A_{tank}}\right) \Delta t$$

where h is the liquid level height, A_{tank} is the cross sectional area of the tank, Δt is the incremental time step, Q_i and Q_o are volumetric flow rates into and out of the tank respectively, and the superscripted values define a particular time step. A tank having a total height of 50 m and a total cross sectional area of 314.13 m^2 (when the gate is up) is simulated. Both the inflow and outflow are allowed to range between 0.0 m^3/sec and 200.0 m^3/sec. A time step of 1.0 sec is used in the discretization of time.

The objective is to develop a FLC capable of driving the liquid level in chamber A, initially set at an arbitrary position with arbitrary initial valve settings, to a given setpoint in the shortest time possible by adjusting the multivalued inflow and outflow. The setpoint is located at the centerline of the tank (h = 25 m). A constraint is placed on the motion of the valves: at no time is either valve allowed to experience an opening or closing that permits a change in flow rate of more than 20 percent of its current setting (maximum change in volumetric flow rate is therefore 40 m^3/sec/sec). This constraint is enforced to limit transients in the piping systems providing flow into and out of the tank. Also, the gate plays an important role in the control problem. The opening and closing of this gate changes A_{tank} by a factor of four, thereby changing the way the liquid level in chamber A responds to changes in the volumetric flow rates. When the gate is allowed to come into play (later in this chapter), an adaptive FLC is needed.

This system is not particularly difficult for a human to control. However, it provides a solid forum for presenting the technique of augmenting FLCs with GAs. This liquid level system can be thought of as a simplified version of more complex environments that have been controlled using FLCs such as flotation cells [30], casting plants [1], and warm water plants [21]. In fact, the FLC described in this chapter has served as the basis for controllers that manipulate complex physical environments [20].

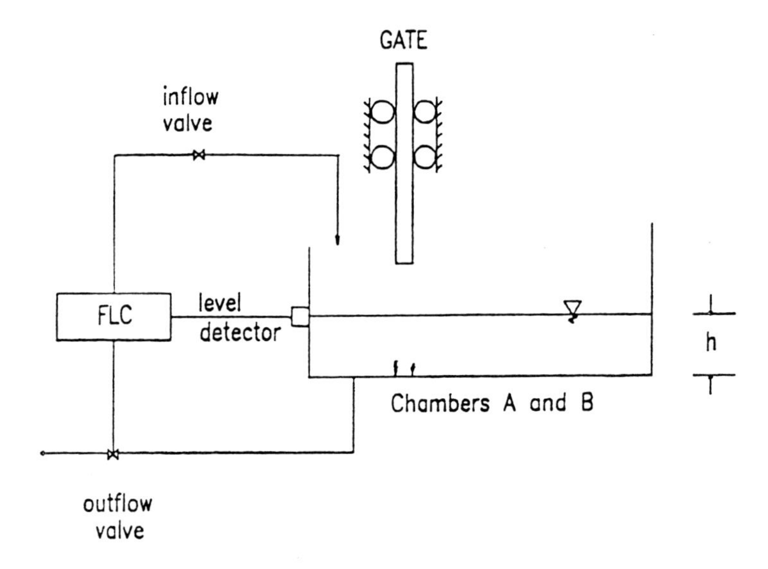

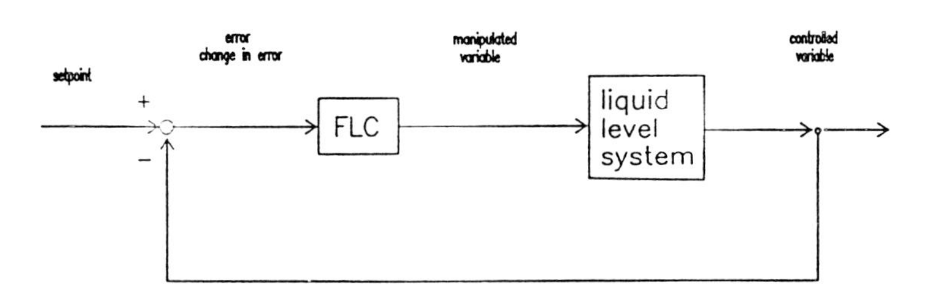

Figure 4.1:[a] The simple liquid level system includes a gate that divides the tank into two chambers, A and B, of equal volume. [b] The block diagram depicting the control loop demonstrates the use of a FLC to drive the liquid level height to a given setpoint.

4.4 DESIGN OF A FUZZY LOGIC CONTROLLER

There are numerous approaches to developing FLCs. Unfortunately, a large number of these approaches are complex and utilize cumbersome *fuzzy mathematics* in the form of a *compositional rule of inference*. In this chapter, a straightforward approach to the development of a FLC is presented that simplifies the application of a compositional rule of inference. A step-by-step procedure for fuzzy control of the liquid level system is provided. This procedure is written in a general form so that it may be easily adapted to the development of other FLCs. In discussing this procedure, special care is taken to relate the mechanics of the FLC to the process a human might use to control the liquid level system. Initially, the gate that divides the tank into two chambers is forced to remain down. The first step in developing the liquid level FLC is to determine which variables will be important in choosing an effective control action. In the artificial intelligence (AI) community these variables are known as the *condition variables*. Determining these variables also happens to be the first thing a human would do in attempting to control a process, and is the initial step in the design of any controller. Two condition variables are readily identified as being important in the liquid level system. First, the current height of the liquid in chamber A, h, is important because it is this height that the FLC must eventually drive to the setpoint. Second, the time rate of change of the height, dh/dt, is important because it describes the rate at which the liquid level is rising or falling. This rate becomes critical as h approaches the setpoint because the constraints placed on the motion of the valves forces the controller to gradually bring the liquid level to the setpoint. In the controls community a controlled variable is the output of the controlled system. One of the condition variables, h, is the control variable. The two inputs to the controller are the error and the change in error defined as:

$$E = h - s$$

$$\Delta E = \frac{dh}{dt}$$

where s is the desired, fixed setpoint.

Once the controlled variable has been chosen, the *manipulated variable* must be identified. Manipulated variables are known in the AI community as the *action variables*. In this liquid level system identifying the manipulated variable is a rather straightforward task for there are only two things the controller can adjust to alter the state of the liquid level system: either the input or output flow rates (Q_i or Q_o) can be increased or decreased by adjusting a valve. To further simplify the problem, Q_i and Q_o were considered together as Q_{net} where:

$$Q_{net} = Q_i - Q_o.$$

In more complex systems, determining the important manipulated variables is not always a straightforward task for there are often a large number of potentially important parameters that have complex relationships with the problem environment.

Once the important controlled and manipulated variables have been identified, the linguistic terms that will be used to describe these variables must be defined (the ***fuzzy sets***). In a classic AI approach to FLC development, fuzzy sets are written to describe the condition variables, h and dh/dt. However, in this discussion a more traditional controls approach is taken. Therefore, fuzzy sets are written to describe E and ΔE. It must be understood that the greater the number of linguistic terms used to describe a variable, potentially the greater the control that can be exerted. However, the more linguistic terms used to describe the condition variables (or in this controls approach, the terms used to describe E and ΔE), the more complex the FLC becomes. Therefore, choosing the number of linguistic terms used to describe each variable is not an arbitrary decision; it is based on knowledge gained through working with the physical system. For the liquid level system, based on the author's experience with the liquid level problem, four fuzzy sets were used to characterize E: **NEGATIVE-BIG (NB)**, **NEGATIVE-SMALL (NS)**, **POSITIVE-SMALL (PS)**, and **POSITIVE-BIG (PB)**. Five fuzzy sets were used to characterize ΔE: **NEGATIVE-BIG (NB)**, **NEGATIVE-SMALL (NS)**, **NEAR-ZERO (NZ)**, **POSITIVE-SMALL (PS)**, and **POSITIVE-BIG (PB)**. Five fuzzy sets were used to characterize Q_{net}: **BIG-POSITIVE (BP)**, **SMALL-POSITIVE (SP)**, **NO-CHANGE (NC)**, **SMALL-NEGATIVE (SN)**, and **BIG-NEGATIVE (BN)**. These fuzzy sets were chosen because they are similar to the descriptive terms a human operator might use to control the liquid level system. The number of linguistic terms used to describe each variable allowed for a FLC of "reasonable" size while providing "adequate" control of the liquid level system.

This choice of fuzzy sets allows for the possibility of 20 different *production rules* to describe all of the possible conditions that could exist in the liquid level system when the rules are of the form:

$$\text{IF [E is } \boldsymbol{A} \text{ and } \Delta\text{E is } \boldsymbol{B}\text{] THEN [}Q_{net}\text{ is } \boldsymbol{C}\text{]}$$

where $\boldsymbol{A}$, $\boldsymbol{B}$, and $\boldsymbol{C}$ are fuzzy sets characterizing the respective variables.

The driving force behind an FLC is the idea that some uncertainty exists in categorizing the values of the system variables. In other words, the linguistic variables mean different things to different people. As a result, there must exist some mechanism for interpreting the fuzzy sets. This mechanism is the *fuzzy membership function*. The fuzzy membership functions used in the liquid level FLC are shown in Figures 4.2, 4.3, and 4.4. Fuzzy membership functions allow the crisp values of E and ΔE to be transformed into a fuzzy set value and the fuzzy control actions of the production rules to be transformed to crisp, discrete control actions. Actually, the fuzzy membership functions can be thought of as determining a degree of confidence in the supposition that a discrete value of a variable is accurately described by a particular fuzzy term; the fuzzy membership function value for E, μ_E, represents the degree to which E is described by a particular fuzzy term. For example, using the membership functions shown in Figure 4.2, a value of E=5 would produce the following membership values: $\mu_{NB}(E=5)=0.0$, $\mu_{NS}(E=5)=0.0$, $\mu_{PS}(E=5)=0.8$, and $\mu_{PB}(E=5)=0.3$. When a fuzzy membership function has a value of $\mu=1.0$, the crisp value is accurately described by the fuzzy term. On the other hand, when $\mu=0.0$, the crisp value is definitely not described by the fuzzy term. It is important to realize that for each crisp control value, each fuzzy set has a membership function value, even though some of the values are zero.

Now that the crisp conditions (definite values of E and ΔE) existing in the liquid level system at any given time can be categorized in a fuzzy set with some certainty, a process for determining a crisp action to take on the liquid level system must be developed. This process involves a set of fuzzy production rules. The set of fuzzy production rules provides a fuzzy action for any condition that could possibly exist in the problem environment. With the four fuzzy sets for E and the five fuzzy sets for ΔE, there exist 20 possible conditions in the liquid level system, and a human expert provides a desirable action for each condition based on prior experience. An example of a fuzzy production rule used in the liquid level system follows:

IF [E is **PB** and ΔE is **PB**] THEN [Q_{net} is **BN**].

This rule simply says that if the liquid level is well above the setpoint and rising rapidly, the net flow into the tank should be made negative big. The complete rule set for the liquid level FLC is depicted in Figure 4.5. This rule matrix is used by locating the descriptive term for E along the top of the matrix, locating the descriptive term for ΔE along the left side of the matrix, and then extracting the appropriate value for Q_{net}.

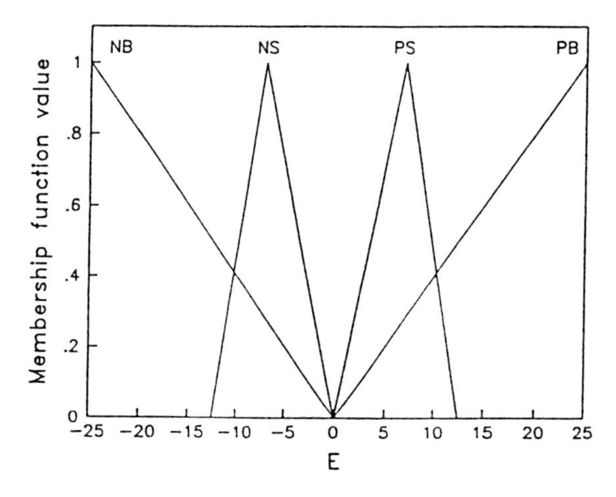

Figure 4.2: Four fuzzy sets were used to describe error.

At this point a means for converting a crisp set of conditions existing in the liquid level system to a set of fuzzy conditions, and a set of fuzzy production rules prescribing a fuzzy action associated with a particular set of fuzzy conditions have been developed. Unlike in conventional expert systems where only one rule is eligible to take effect, all of the rules in a FLC take effect to some degree at every time step. Therefore, there still remains the task of converting the 20 fuzzy actions prescribed by the fuzzy production rules into a single, crisp action to be taken on the liquid level system. Larkin [24] found that a center of area (COA) method (sometimes called the centroid method) provides an efficient means for determining this crisp action. In the COA method, the fuzzy membership functions shown in Figure 4.4 are used in a weighted summing procedure to find one crisp action. These manipulated variable membership functions actually serve only to define the base points of the triangles because the heights of the membership functions are varied according to the membership function values associated with the individual rules as applied to a particular situation. Each production rule has its action (as described by a fuzzy membership function) re-plotted with a height equal to the minimum likelihood associated with the condition portion of the rule. For example, consider the sample rule provided above as applied when the conditions in the system are E=5 and ΔE=0.521. Since $\mu_{PB}(E=5)=0.3$ and $\mu_{PB}(\Delta E=0.521)=0.7$, the membership function for Q_{net}=**BN** is re-plotted with a height of 0.3. This is because the minimum membership function value for the variables on the condition portion of the rule is 0.3. When all 20 conditions have been compared to the rules, 20 triangles of varying height will have been re-plotted. Figure 4.6 shows a ***solution plot*** for the aforementioned condition existing in the environment: E=5 and ΔE=0.521. The two triangles occur as a consequence of the two rules that have their conditions met to some degree (the other 18 rules have a portion of their conditions met to the degree of 0.0). The crisp action is determined simply by finding the center of area of the total area comprised of the 20 triangles; in this case the crisp action is Q_{net}=-12.0 percent. The center of area defines the single value of percent change in Q applied at the next time step. The rules that create triangles with large heights, those arising from the conditions that are most certain, have the greatest effect on the action.

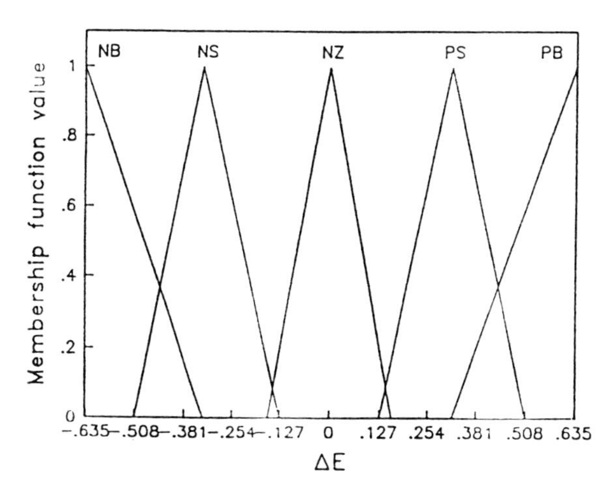

Figure 4.3: Five fuzzy sets were used to describe time rate of change of error.

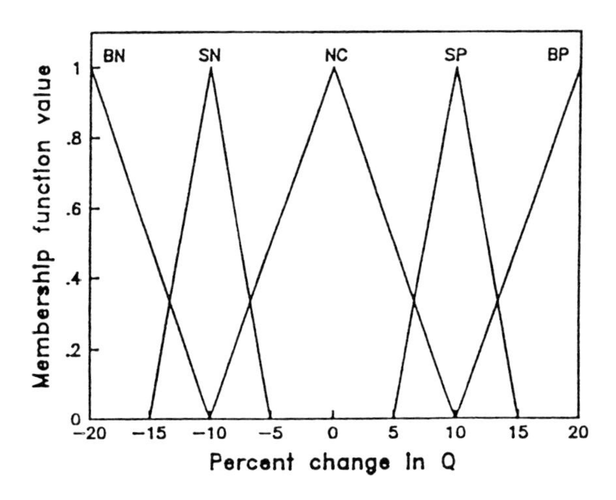

Figure 4.4: Five fuzzy sets were used to describe percent change in flow rate.

E

ΔE	BN	SN	SP	BP
BN	BP	BP	SP	SN
SN	BP	SP	NC	SN
NZ	BP	SP	SN	BN
SP	SP	NC	SN	BN
BP	SP	SN	BN	BN

Figure 4.5: The rule set included 20 rules initially written by the author and later changed by a micro GA.

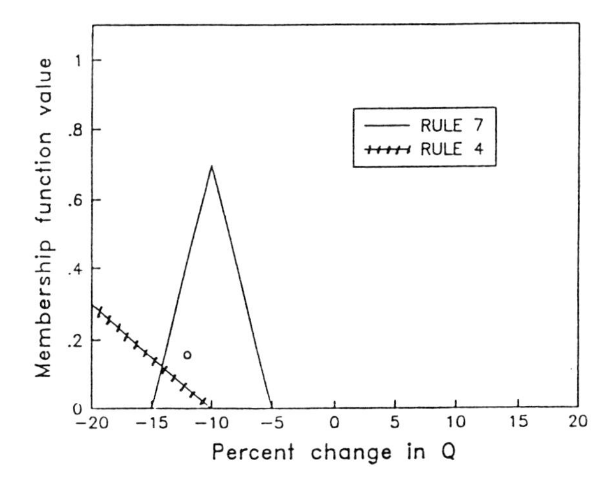

Figure 4.6: The center of area method (sometimes called the centroid method) is a commonly used technique for defuzzification.

The step-by-step procedure described above for developing an FLC is summarized below:

1) Determine the controlled variables to be considered;

2) Determine the manipulated variables to be considered;

3) Describe the fuzzy sets for both the controlled and manipulated variables (in a traditional controls approach error and change in error are sometimes describe instead of the controlled variables);

4) Establish a set of fuzzy production rules that cover all of the possible conditions that could exist in the problem environment;
5) Define the fuzzy membership functions;

6) Compare the set of conditions existing in the problem environment to the production rules, and compute the desirable value of Q_{net} using the rule matrix appearing in Figure 4.5.

Certainly, this procedure can be easily put in the form of a computer program.

To someone who has never been exposed to FLCs, this may seem like a dubious approach to process control. To someone who has worked extensively with FLCs (and possibly even to those who have progressed through this book to Chapter 5), this approach hopefully seems logical and quite possibly natural. At any rate, to help the reader become more comfortable with the effectiveness of this *fuzzy approach* to process control, consider a FLC computer program that manipulates a mathematical model of the liquid level system. Figure 4.7 shows the liquid level height plotted as a function of time for one particular initial condition for the problem environment. The FLC uses only the set of 20 fuzzy production rules to govern its selection of actions. The FLC is able to drive the height to the setpoint of 25 m in approximately 70 sec at which time a small oscillation about the setpoint occurs.

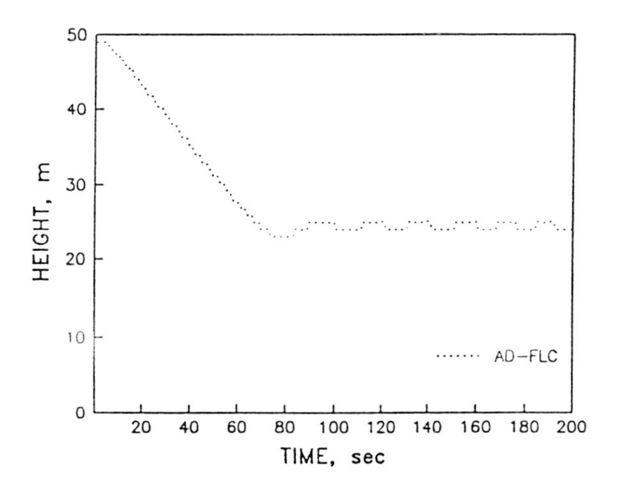

Figure 4.7: The author-developed FLC successfully drives the liquid level to the setpoint.

This section has presented a straightforward approach to the development of a FLC. Since the details of the FLC have been provided (including the membership functions, the rule set, and the defuzzification scheme), the reader should be able to reproduce the results presented. Although the steps in this development for the liquid level system are conceptually simple, two of the necessary steps can indeed be frustrating. The selection of both the rule set and the membership functions can present the FLC developer with some intriguing decisions. In the remainder of this chapter, a particular small-population GA, a *micro GA*, will be used to accomplish these two tasks. However, it is important to note that a micro GA was simply the author's "GA of choice." More traditional GAs (like the ones discussed in earlier chapters), should provide the same results. Before the details of the application of a micro GA to the tasks of learning rules and membership functions are discussed, the mechanics of a micro GA are presented.

4.5 THE MECHANICS OF A MICRO GA

A micro GA [19, 23] is a particular GA that uses small populations in an attempt to improve the speed with which GAs locate near-optimal solutions. This scheme reduces the number of function evaluations required to solve large-scale search problems by reducing the number of function evaluations required in each generation. Actually, the approach employed by a micro GA is contrary to traditional GA strategies. To appreciate this point, one must consider a piece of GA history.

Historically, the choice of parameters used in GA applications (population size, reproduction probability, mutation probability, etc.,) has been based on suggestions made in studies by De Jong [5] and Grefenstette [11]. Both of these studies indicate "bigger is better" when selecting a population size. These researchers brought to light the fact that when population sizes are too small, premature convergence can occur—the GA rapidly converges to a sub-optimal solution. Because of this trend, population sizes in conventional GA applications are normally on the order of the length of the bit-strings used by the GA. These large populations allow for convergence to near-optimal solutions, but often require an excessive number of function evaluations when solving problems with a substantial number of parameters. Having to make numerous function evaluations is not of great concern when either the function evaluations are inexpensive, or when time to convergence is not at a premium. Unfortunately, neither of these conditions apply when learning membership functions for FLCs. Although when learning membership functions in the initial design phase, the time to convergence is not of great concern because large population GAs locate membership functions that are more effective than those determined by humans via trial-and-error [1]. And, at this stage of development, search time is not at a premium.

In a recent work, Goldberg [8] describes an approach for the successful use of small populations in GA applications. This approach, implemented in the micro GA, helps the GA overcome shortcomings apparent in small population GAs (those of limited information processing and premature convergence). In fact, the micro GA actually uses premature convergence to its advantage. It forces strings to converge quite rapidly, and then uses the information contained in the converged solution to its fullest advantage.

Now that the micro GA has been presented in the context of GA history, the details of a micro GA application can be discussed. Possibly the best way to understand the mechanics of the micro GA is to consider an application to a specific problem. The application of a micro GA to an adaptive FLC is, of course, presented in this chapter. However, it is prudent to consider a simpler application to first get a feel for the behavior of this particular form of GA.

Consider a straightforward search problem; consider the problem of maximizing the function $g(x)=x^2$ for the integer values of x such that $0 \leq x \leq 7$. The function is plotted in Figure 4.8. It should be readily apparent that the solution to the search problem is $g(x)=49$ at the value of $x=7$; this is a simple problem. The purpose of presenting this example is to prevent the details of a micro GA application from getting clouded by the complexity of the search problem.

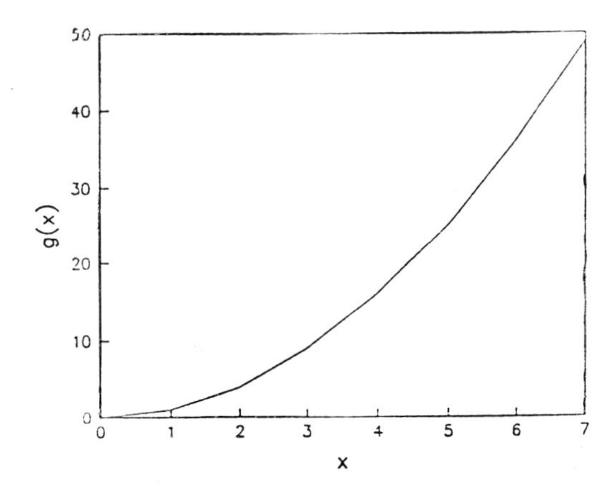

Figure 4.8: The mechanics of a micro GA are demonstrated above.

Before the details of the micro GA are presented, two initial decisions must be made: (1) how to code the possible solutions to the search problem as bit-strings and (2) how to evaluate the merit of the possible solutions. For this example problem, a popular coding approach will be adopted. Although there are a number of effective coding schemes [9], the solutions to the membership function search problem will be represented with binary numbers as follows:

bit-string	x value
000	0
001	1
010	2
011	3
100	4
101	5
110	6
111	7

The merit of the possible solutions is represented by a fitness function, f. Traditionally, the higher the value of f, the better the solution. A common-sense fitness function for this problem is f=g(x). This way, those strings that produce the highest values of g(x), have the highest fitness values.

Now that the preliminary decisions concerning coding and fitness have been made, the mechanics of the micro GA can be discussed. The basic approach to implementing a micro GA are summarized in the following steps (where a specific example to the maximization problem has been included):

1) Generate 5 strings randomly. A randomly generated initial population might be composed of the following strings:

string#	string #	x value	f
1	000	0	0
2	010	2	4
3	100	4	16
4	001	1	1
5	100	4	16

Subsequent generations (see step 6) are produced by applying the traditional mutation operator to the current best string. For instance, if the current best string is 101, the four new strings needed to complete the population would be slight variations of the current best. For instance, 100, 111, 100, and 001 would be possibilities.

2) Rank the strings in the current population according to fitness. For the initial generation, strings 3 and 5 have the largest fitness values, and would thus be ranked 1 and 2. String 1 has the smallest fitness value and thus would be ranked 5. After ranking, the initial generation would appear as follows:

string #	string	x value	f
3	100	4	16
5	100	4	16
2	010	2	4
1	001	1	1
4	000	0	0

3) Save a copy of the best string for possible use in step 6.

4) Perform reproduction (empirical studies have shown that tournament selection [19] works well). For tournament selection, the string with the highest fitness between adjacent string pairs (i.e., 3 to 5, 5 to 2, 2 to 1, 1 to 4, 4 back to 3) is copied into a mating pool. For the example being discussed, string number 3 is compared to string number 5. They both have the same fitness value, so either one is copied into the mating pool. String number 5 is then compared to string number 2. String number 5 has the higher fitness value, so it is copied into the mating pool. When all of the comparisons have been made, the mating pool consists of the following strings:

string #	string #	x value	f
3	100	4	16
5	100	4	16
2	010	2	4
1	001	1	1
3	100	4	16

5) Perform conventional single-point crossover. Crossover has a random element so the population can have a number of configurations after

crossover. However, one possible configuration of the population after crossover for the example being discussed follows:

string #	string	parents	x value	f
1	110	3, 2	6	36
2	101	1, 5	5	25
3	110	2, 5	6	36
4	101	3, 1	5	25
5	011	2, 1	3	9

6) Make sure the best string in the current population is at least as good as the string that was saved in step 3. If not, remove the string with the smallest fitness, and replace it with the string from step 3.

7) Test population for convergence. In this example, convergence is defined using Hamming distance; that is, when all population members are the same in all but N bit positions. Here string length is 3, a reasonable value for N is 0 or 1.

8) If population has converged, select the string with highest fitness value and use it to generate a new population. This new population is generated using an arbitrary criterion. Suppose the best string from step 7 was 101, the other four strings might be generated by randomly complementing a single bit position. Possible results might be 100, 111, 100, and 001. Take this population and go to step 2.

This iterative process is continued until a final termination condition is met such as a time limit, a maximum number of function evaluations, or the optimum has been found.

Traditionally, GAs have attempted to avoid premature convergence at all costs. In contrast, the micro GA encourages convergence. A micro GA attempts to get the absolute most it can out of the encoded information appearing in its small population by rapidly converging to a form of the best string in the current population. Then, this small population GA uses strings that are similar to the current best solution to further its search. In other words, it attempts to use the genetic-like material of the current best string to guide its search. Thus, a micro GA turns what has traditionally been thought of as a drawback to small populations into an advantage.

4.6 THE DESIGN OF A NON-ADAPTIVE LIQUID LEVEL FLC USING A MICRO GA

In this section, a micro GA is used to design a non-adaptive FLC for the liquid level system (the definition of adaptive FLC as it pertains to this discussion is provided in the next section). Initially, a micro GA chooses a rule set. Membership functions for this rule set are chosen to represent the author's conception of the fuzzy terms needed to describe E, ΔE, and Q_{net}. Next, this rule set is held fixed, and a micro GA adjusts the membership functions until the FLC achieves a satisfactory performance level. In both these cases, the GA's search is a part of the initial design phase, and the rules and membership functions are not changed once the FLC is connected to the liquid level system.

There are basically two decisions to be made when applying a GA to a search problem: (1) how to code the possible solutions to the problem as finite bit strings and (2) how to evaluate the merit of each string. In this problem, the parameters to be coded are the action portions of the 20 rules used previously. It has been reported that binary codings (the use of bit strings) produce the most efficient genetic searches [9]. Due to this, and because of the nature of the parameters involved in the search for efficient rules, a direct mapping is used. Specifically, each of the 20 rules are represented by a three bit parameter in a 60 bit string of ones and zeros. The first three bits represent the action portion of the first rule, the second three bits represent the action portion of the second rule. This pattern continues until the final three bits represent the action portion of the 20th rule. The coding for the individual parameters is summarized in Table 4.1.

Bit representation	Parameter
000	Big Negative
001	Big Negative
010	Small Negative
011	No Change
100	No Change
101	Small Positive
110	Big Positive
111	Big Positive

Table 4.1: The coding used to select an efficient rule set required a 3-bit string.

Now that an appropriate coding has been determined, the issue of evaluating the merit of each string (each rule set) must be addressed. The task of defining a fitness function is always application specific; it always comes down to accurately describing the goal of the controller. In the case of the liquid level FLC, the objective is to drive the level in chamber A to a specified setpoint as rapidly as possible, and to keep the level at the setpoint once it has been reached. The ability of the FLC to achieve these objectives can be represented by a fitness function that specifies how well the controller reduces the error over some finite time period. Additionally, to ensure a robust FLC is obtained, the fitness function should reflect the controller's ability to reach the setpoint from a number of initial condition cases. Mathematically, this fitness function is expressed as:

$$f = \sum_{i=case\ 1}^{case\ 4} \sum_{j=0\ sec}^{30\ sec} E^2.$$

The initial condition cases are selected from different regions of the state space of the liquid level system. The four initial condition cases are:

case #	h	dh/dt
1	0.0	0
2	50.0	0
3	12.5	-0.635
4	37.5	0.635

Using the aforementioned coding and fitness function, the non-adaptive FLC is able to determine an action for each of the 20 rules that allows the FLC to drive the liquid level to the setpoint. The micro GA is able to locate the efficient rule-set developed by the author (the human *expert*). The resulting FLC demonstrated the same characteristics as the AD-FLC whose performance is shown in Figure 4.7. However, a troublesome question remains: are the membership functions selected by the author the most effective? The answer is: it is highly unlikely. And, in fact, adjusting the membership functions slightly can improve the FLC's ability to drive the liquid level to the setpoint.

When using a micro GA in the quest for efficient membership functions, the same two decisions must be made that had to be made when using a GA to locate an efficient rule set. For the purpose of this example, the membership functions maintain a triangular shape, but were allowed to change position relative to each other. The triangles describing ***extreme fuzzy sets*** (those at the extremes of the values they were describing, terms like **BP** and **BN**, which formed right triangles) were forced to remain fixed to their associated extreme limits. Membership functions defining ***interior fuzzy sets*** were forced to remain isosceles triangles. Thus, the search for efficient membership functions entailed 22 parameters (22 points are needed to locate the 14 triangles that described the fuzzy terms). The 22 parameters that were needed to completely describe the membership functions consisted of the base points for each of the triangles that define the membership functions. Each extreme fuzzy set required one parameter to completely define it, while each interior fuzzy set required two parameters to completely define it.

A common coding called ***concatenated, mapped, unsigned binary coding*** was used in this application. A variable, the right most point defining the limit on the membership function of E=**NB** for example, was discretized by mapping from a minimum value C_{min}=-25 to a maximum value C_{max}=25 using a four-bit, unsigned binary integer according to the equation:

$$C = C_{min} + \frac{b}{(2^l - 1)} * (C_{max} - C_{min})$$

where b is the integer value represented by an l-bit string and C is the value of the parameter being coded. As in the coding of the rule set, the binary values for the different parameters are concatenated into a single bit-string. Thus, each string represents an entire set of fuzzy membership functions. Since the objective of the FLC is the same as it was in the previous example of selecting a rule set, the same fitness function can be used.

A micro GA is used to select the membership functions for the liquid level FLC that used the rule set located previously using a GA. This time, the rule set was held fixed while the membership functions were altered. After having sampled a small portion of the search space (approximately 3200 of the $2^{88}=3.09*10^{26}$ possible points), the micro GA was able to select membership functions that provided for better control than those defined by the author. Figures 4.9 and 4.10 represent comparisons of the GA-FLC to the liquid level FLC developed by the author (AD-FLC) for two sets of initial conditions (one of these sets of initial conditions was used in the definition of the fitness function, and one of these sets of initial conditions did not appear in the definition of the fitness function). In each case, the GA-FLC drove the liquid level to the setpoint faster and maintained the setpoint better than the AD-FLC.

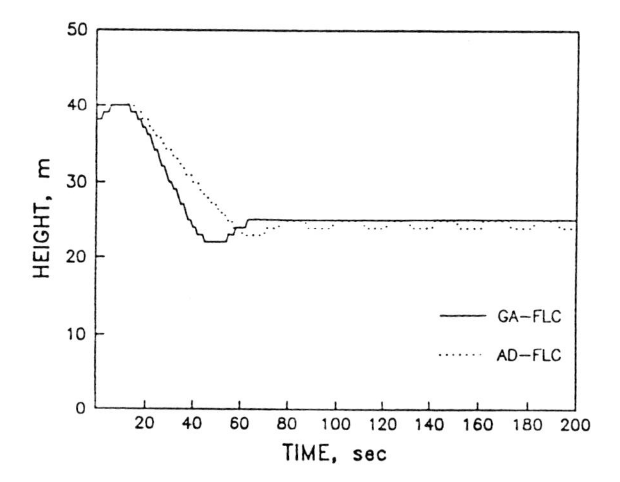

Figure 4.9: The genetic algorithm-developed FLC outperforms the author-developed FLC for a set of initial conditions that were included in the genetic algorithm's fitness function.

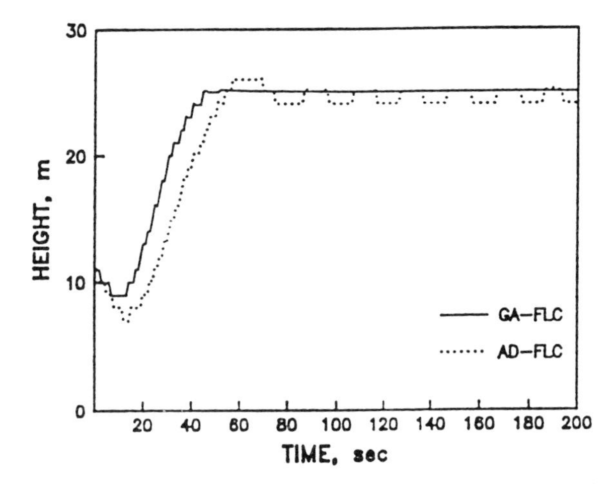

Figure 4.10: The genetic algorithm-developed FLC outperforms the author-developed FLC for a set of initial conditions that were not included in the genetic algorithm's fitness function.

4.7 AN ADAPTIVE FLC FOR A TIME-VARYING ENVIRONMENT

Now that the technique for developing FLCs using GAs has been presented, it is time to consider the previously neglected gate in the liquid level system. When the position of this gate is manipulated by an external agent, the control problem becomes ***time-varying***. When considering the design of a FLC for controlling this time-varying problem, the initial reaction is to simply include the position of the gate as a condition variable. Certainly, this is a viable solution, and the inclusion of two different values of A_{tank} in the rule set would not cause the size of the rule set to become extremely large. However, if a number of different externally controlled variables were introduced, or if a limited number of externally controlled variables required a large number of linguistic terms to describe them, the size of the rule set could quickly become too large to effectively manage. Thus, instead of re-writing the existing rule set to include another condition variable, an alternative approach is considered. Namely, an adaptive FLC is produced. An adaptive controller refers to one that is able to account for changes in variables that do not explicitly appear in the controller's rule set by altering either its membership functions or its rule set "on-line."

It has already been established in this chapter that altering the membership functions used in the liquid level FLC can substantially alter the performance of the controller. By allowing on-line alteration of membership functions, the adaptive FLC is capable of modifying its interpretation of the fuzzy terms in response to changes in the environment it is controlling. The adaptation in real time ia achieved by allowing the FLC to employ the micro GA.

The adaptive FLC receives the current values of E and ΔE for the liquid level system. The FLC then uses a micro GA to explore for membership functions that are capable of accounting for the changes in the response of the liquid level system caused by the opening of the gate. Figure 4.11 shows a schematic of the organization of the adaptive controller. The FLC manipulates the liquid level system while the adaptive element searches for membership functions producing improved FLC performance. Using this basic approach, the micro GA is able to locate high-performance membership functions on-line.

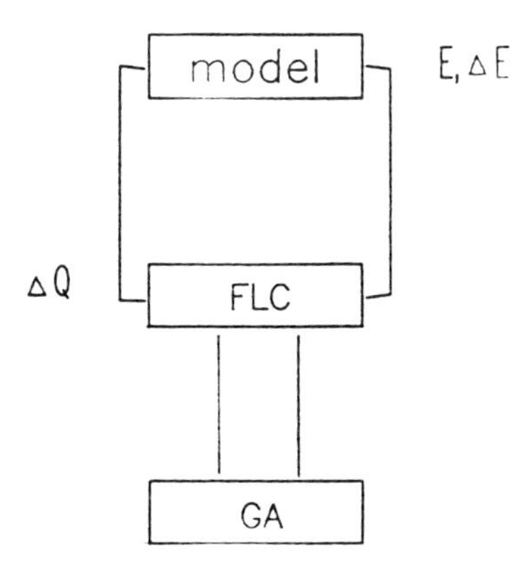

Figure 4.11: A schematic of the adaptive FLC that uses a genetic algorithm to select membership functions "on-line."

Basically, there is nothing new in this section as far as the application of a micro GA to the selection of effective membership functions is concerned. The only difference is that the micro GA is used for exploration while the FLC is actually controlling the environment. Therefore, it suffices to simply present results demonstrating the effectiveness of an adaptive FLC that uses a micro GA to control the liquid level system. These results appear in Figure 4.12. The non-adaptive FLC does not perform as well as the adaptive FLC because its conception of ΔE is based on the original area of the tank. However, when the gate is opened, ΔE no longer reacts the same to changes in flow rate (the relationship changes by a factor of four).

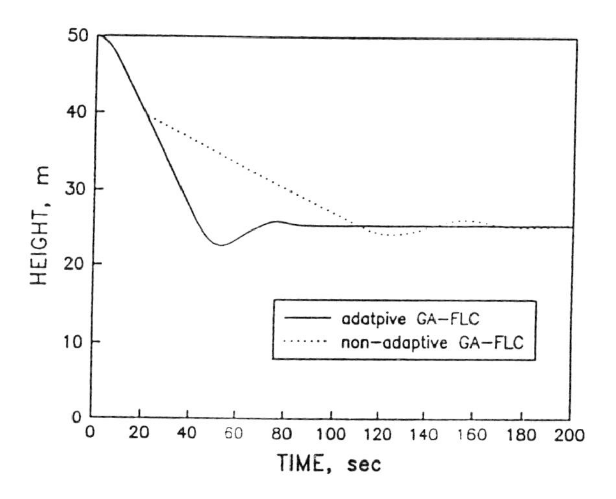

Figure 4.12: The adaptive FLC, augmented with a genetic algorithm, outperforms a non-adaptive FLC because of the changing process dynamics.

This approach to the development of adaptive FLCs certainly has some potential benefits, and is applicable to the control of a number of physical systems (as will be seen in the next chapter). However, there are two limitations to this approach that should be mentioned. First, if the environment being controlled is changing rapidly relative to the time the micro GA needs to locate efficient membership functions, the controller will always be behind. Therefore, rapidly changing systems do not lend themselves to this approach (at least until faster computers or more improved techniques are developed). As a point of reference, it took the micro GA approximately 15 sec to locate membership functions for the liquid level system when running on a 486-based personal computer. Second, there must exist at least a crude model of the system being controlled. The micro GA uses a model of the physical system to evaluate the merit of potential membership function sets while the FLC manipulates the actual physical system. This, of course, can present insurmountable obstacles for some systems.

4.8 SUMMARY AND CONCLUSIONS

Researchers at the Universities of Alabama and South Australia have developed a technique that combines the process control capabilities of fuzzy logic with the search capabilities of GAs. In this chapter, the technique has been presented and applied to a computer simulation of a simple system. Initially, a micro GA was used to select both the rule set and the membership functions for a non-adaptive liquid level FLC. In this initial example of GA-augmented FLC design, the GA-FLC outperformed an AD-FLC. Next, a micro GA was used to provide the original non-adaptive FLC with adaptive capabilities. In this second example of GA-augmented FLC design, the GA selected high-performance membership functions that allowed the fixed rule set to adequately account for changes in the position of a gate that could be used to divide the tank into two chambers.

FLCs have become increasingly viable solutions to process control problems. However, if the utility of these rule-based systems is going to continue to grow, techniques must be developed that reduce FLC development time and ease the task of writing the necessary rule sets. Also, if FLCs are going to provide practical solutions to complex process control problems like those found in the mineral processing industry and waste treatment plants (two areas of prime concern to the U.S. Bureau of Mines), techniques must be developed that will provide FLCs with adaptive capabilities. Based on the results presented in this chapter, it is concluded that GAs represent such a technique and can provide a valuable resource for the design of FLCs.

BIBLIOGRAPHY

[1] Bartolini, G., Casalino, G., Davoli, F., Mastretta, M., Minciardi, R., and Morten, E. (1985). Development of performance adaptive fuzzy controllers with application to continuous casting plants, *Industrial Applications of Fuzzy Control* (M. Sugeno, ed.), North-Holland, Amsterdam, pp. 73-85.

[2] Booker, L. B. (1982). "Intelligent Behavior as an Adaptation to the Task Environment", *Dissertation Abstracts International*, vol. 43, no. 2, p. 469B.

[3] Davis, L. (1985). "Job Shop Scheduling With Genetic Algorithms," Proceedings of an International Conference on Genetic Algorithms and their Applications, Boston, MA, pp. 136-140.

[4] Davis, L., and Smith, D. (1985). "Adaptive Design for Layout Synthesis," internal report, Texas Instruments, Dallas, TX.

[5] De Jong, K. A. (1975). "Analysis of the Behavior of a Class of Genetic Adaptive Systems," Ph. D. Thesis, University of Michigan, Ann Arbor, MI.

[6] Evans, G. W., Karwowski, W., and Wilhelm, M. R. (1989). *Applications of Fuzzy Set Methodologies in Industrial Engineering*, Elsevier, Amsterdam.

[7] Fuzzy Logic'93 (1993). Conference Proceedings, Buoilinganne, Published by Computer Design Magazine, a Penn Well Publication.

[8] Goldberg, D. E. (1988). "Sizing Populations for Serial and Parallel Genetic Algorithms," Report Number 88004, The Clearinghouse for Genetic Algorithms, Tuscaloosa, AL.

[9] Goldberg, D. E. (1989). *Genetic Algorithms in Search, Optimization, and Machine Learning*, Addison-Wesley, Reading, MA.

[10] Goldberg, D. E., and Samtani, M. P. (1986). "Engineering Optimization Via Genetic Algorithm," Proceedings of the Ninth Conference on Electronic Computation, Birmingham, AL, pp. 471-482.

[11] Grefenstette, J. J. (1986). "Optimization of Control Parameters for Genetic Algorithms", *IEEE Transactions on Systems, Man, and Cybernetics*, Vol. SMC-16, pp. 122-128.

[12] Holland, J. H. (1975). *Adaptation in Natural and Artificial Systems*, The University of Michigan Press, Ann Arbor, MI.

[13] Holland, J. H., and Reitman, J. S. (1978). "Cognitive Systems Based on Adaptive Algorithms", *Pattern Directed Inference Systems* (D. Waterman, and R. Hayes-Roth, eds.), Academic Press, New York, NY.

[14] Hollstien, R. B. (1971). "Artificial Genetic Adaptation in Computer Control Systems," Ph. D. Thesis, University of Michigan, Ann Arbor, MI.

[15] Jain, L. C., and Karr, C. L. (1995). Introduction to Fuzzy Systems, Chapter 3, ETD2000, IEEE Computer Society Press, USA.

[16] Jain, L. C., and Karr, C. L. (1995). Introduction to Evolutionary Computing Techniques, Chapter 4, ETD2000, IEEE Computer Society Press, USA.

[17] Kandel, A., and Langholz, G., Editors (1994). *Fuzzy Control Systems*, CRC Press, USA.

[18] Karr, C. L. (1989). "Analysis and Optimization of an Air-Injected Hydrocyclone," Ph. D. Thesis, The University of Alabama, Tuscaloosa, AL.

[19] Karr, C. L. (1991). Genetic Algorithms for Fuzzy Controllers, *AI Expert*, Vol. 6, No. 2, pp. 26-33.

[20] Karr, C. L., and Gentry, E. J. (1992). "Real-Time pH Control Using Fuzzy Logic and Genetic Algorithms," Proceedings of Annual Meeting of the Society for Mining, Metallurgy, and Exploration, Phoenix, AZ, February, 1992.

[21] Kickert, W. J. M., and Van Nauta Lemke, J. R. (1976). "Applications of a Fuzzy Controller to a Warm Water Plant", *Automatica*, Vol. 12, pp. 301-308.

[22] Koza, J. R. (1994). *Genetic Programming II*, The MIT Press, USA.

[23] Krishnakumar, K. (1991). "Fuzzy Genetic Algorithms," Proceedings of a Workshop on Neural Networks, pp. 31-37.

[24] Larkin, L. I. (1985). "A Fuzzy Logic Controller for Aircraft Flight Control", *Industrial Applications of Fuzzy Control* (M. Sugeno, ed.), North-Holland, Amsterdam, pp. 87-96.

[25] Lee, M. A., and Takagi, H. (1993). "Integrating Design Stages of Fuzzy Systems Using Genetic Algorithms", Proceedings of the Second International Conference on Fuzzy Systems (FUZ-IEEE'93), vol. 1, pp. 612-617.

[26] Lee, M. A., and Takagi, H. (1993). "Embedding A priori Knowledge into an Integrated Fuzzy System Design Method Based on Genetic Algorithms", Proceedings of the 5th IFSA World Congress (IFSA'93), vol. II, pp. 1293-1296.

[27] Michalewicz, Z. (1994). *Genetic Algorithms + Data Structures = Evolution Programs*, Springer-Verlag, USA.

[28] Procyk, T. J., and Mamdani, E. H. (1978). "A Linguistic Self-Organizing Process Controller", *Automatica*, Vol. 15, pp. 15-30.

[29] Ross, T. J. (1995). *Fuzzy Logic with Engineering Applications*, McGraw-Hill, Inc., USA.

[30] Shah, I., and Rajamani, K. (1988). "Fuzzy Logic Controller: Application to Liquid Level System," Proceedings of IFAC Symposium on Automation in Mining, Metallurgy and Metals Processing, Johannesburg, pp. 186-192.

[31] Sugeno, M., Editor (1985). *Industrial Applications of Fuzzy Control*, Elsevier, Amsterdam.

[32] Waterman, D. A. (1989). *A Guide to Expert Systems*, Addison-Wesley, Reading, MA.

[33] Zadeh, L. A. (1965). Fuzzy Sets, *Information and Control*, Vol. 8, pp. 338-369.

Chapter 5

Cases in geno-fuzzy control

Charles L. KARR
Department of Aerospace Engineering & Mechanics
University of Alabama, Box 870280
Tuscaloosa, AL 35487-0280, USA
ckarr@eng.ua.edu

Lakhmi C. JAIN
Knowledge-Based Intelligent Engineering Systems
University of South Australia
Adelaide, The Levels, SA, 5095, Australia
etlcj@levels.unisa.edu.au

Researchers at the Universities of Alabama and South Australia have developed a technique that utilizes the search capabilities of genetic algorithms to enhance the process control capabilities of fuzzy logic. This chapter describes several applications of the technique.

5.1 INTRODUCTION

In the previous chapter a technique was introduced in which the search capabilities of genetic algorithms (GAs) are used to enhance the process control capabilities of fuzzy logic. A GA-enhanced fuzzy logic controller (FLC) for a liquid level system was used to demonstrate the efficacy of the method. Admittedly, the liquid level system is easily controlled by a human operator, and thus the true power of combining GAs with FLCs may not have been conveyed to the reader. Therefore, in the current chapter, the technique is applied to two control problems that are more difficult to manage: a computer-simulated cart-pole balancing system and a laboratory pH titration. Additionally, a GA is used to control a backpropogation neural network.

The two control problems were selected for a number of reasons. First, both of the systems considered are nonlinear and therefore represent difficult control problems. It would be both a difficult and a time-consuming task for a human to learn to efficiently manipulate the systems. Second, both of the systems are representative of problems that must be tackled in a number of today's industries. The cart-pole balancing system is often used as a model of difficult balancing systems, while the control of pH is vital to the mineral processing industry. Third, both of these problems have received attention in the literature [1, 2, 3, 10, 12], and consequently, there are studies against which the performance of the GA-FLCs can be compared.

5.2 CART-POLE BALANCING SYSTEM

The problem of interest in this application is the control of a cart-pole balancing system. A cart is free to translate along a one-dimensional track while a pole is free to rotate only in the vertical plane of the cart and track. A multivalued force, F, can be applied at discrete time intervals in either direction to the center of mass of the cart. A schematic of the cart-pole system is shown in Figure 5.1a. The objective of the control problem is to apply forces to the cart until it is motionless at the center of the track and the pole is balanced in a vertical position. A block diagram of the control loop is shown in Figure 5.1b. This task of centering a cart on a track while balancing a pole is often used as an example of the inherently unstable, multiple-output, dynamic systems present in many balancing situations, e.g., two-legged walking and the aiming of a rocket thruster.

The state of the cart-pole system at any time is described by four real-valued state variables:

x = position of the cart;
$\dot{x}$ = linear velocity of the cart;
θ = angle of the pole with respect to the vertical;
and $\dot{\theta}$ = angular velocity of the pole.

The system is modeled by the nonlinear ordinary differential equations [2]:

$$\ddot{\theta} = \frac{g \sin\theta + \cos\theta \left[\frac{-F - m_p l \dot{\theta}^2 \sin\theta + \mu_c \, sign(\dot{x})}{(m_c + m_p)}\right] - \frac{\mu_p \dot{\theta}}{m_p l}}{l\left[\frac{4}{3} + \frac{m_p \cos^2\theta}{(m_c + m_p)}\right]} \quad (1)$$

$$\ddot{x} = \frac{F + m_p l \left[\dot{\theta}^2 \sin\theta - \theta \cos\theta\right] - \mu_c \, sign(\dot{x})}{(m_c + m_p)} \quad (2)$$

where:

g	=	$-9.81 \, m/s^2$, acceleration due to gravity;
m_c	=	1.0 kg, mass of cart (as will be discussed later, this value changes with time);
m_p	=	0.1 kg, mass of pole;
l	=	0.5 m, length of pole;
μ_c	=	0.0005, coefficient of friction of cart on track;
μ_p	=	0.000002, coefficient of friction of pole on cart;

and $-10.0 \, N \leq F \leq 10.0 \, N$, force applied to cart's center of mass.

The solution of these equations was approximated using Euler's method, thereby yielding the following difference equations:

$$\dot{\theta}^{t+1} = \dot{\theta}^t + \ddot{\theta}^t \Delta t \quad (3)$$

$$\theta^{t+1} = \theta^t + \dot{\theta}^t \Delta t \quad (4)$$

$$\dot{x}^{t+1} = \dot{x}^t + \ddot{x}^t \Delta t \quad (5)$$

$$x^{t+1} = x^t + \dot{x}^t \Delta t \quad (6)$$

where the superscripts indicate values at a particular time, Δt is the time step, and the values x^t and θ^t are evaluated using equations (1) and (2). A time step of 0.02 seconds was used because this time step struck a balance between the accuracy of the solution and the computational time required to find the solution. As will be seen later, the GA's ability to locate effective membership functions depends heavily on the computational time required for a simulation.

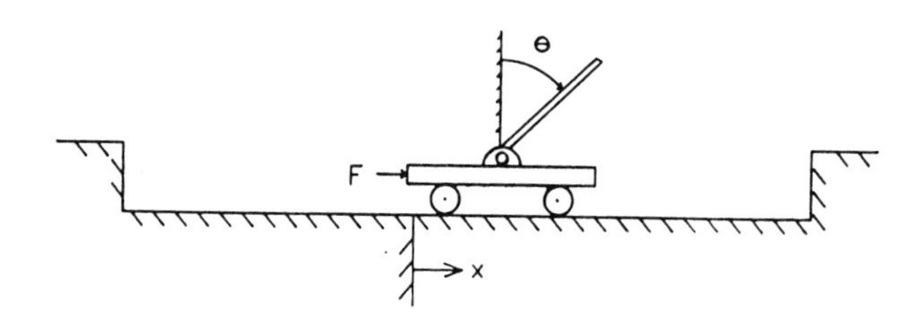

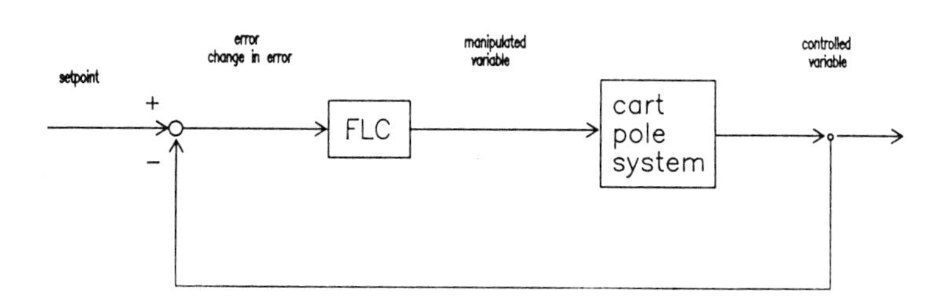

Figure 5.1: The cart-pole balancing system is a popular test for control strategies. A schematic of the system is shown in [a] while a control loop for the system is shown in [b].

The preceding is a description of the classic cart-pole system as it is generally addressed in the literature. The characteristic parameters of the physical system, parameters such as cart mass and pole length, remain constant in the problem traditionally solved. However, in this paper, the control problem is made considerably more difficult by considering a system in which the mass of the cart changes with time as shown in Figure 5.2. This expansion transforms the problem into a time-varying control problem. Note that the cart mass increases by 500 percent which significantly alters the response of the cart-pole system to a given force stimulus. To keep the size of the rule set required by the FLC to a minimum and to reduce the computation time needed by the FLC to select an appropriate action, the mass of the cart is not included in the rule set. Therefore, changes in the response of the cart-pole system must be accounted for by altering the membership functions in real-time (or by altering the rule set, an alternative that is not investigated in this presentation). Thus, an adaptive FLC is required: one that is able to account for changes in variables that do not explicitly appear in the controller's rule set.

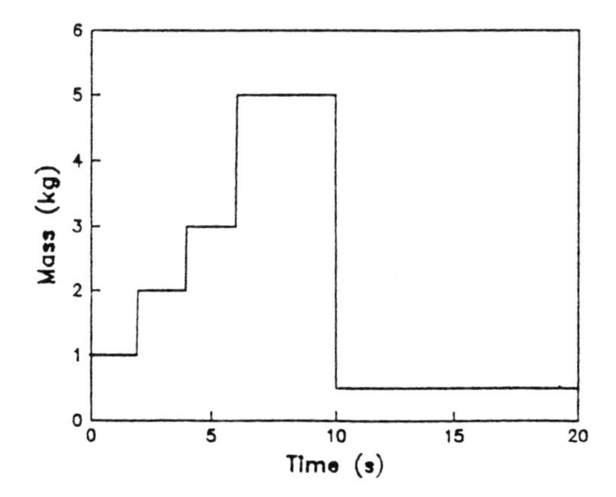

Figure 5.2: The FLC is forced to maintain control of the cart-pole system despite harsh step changes in the mass of the cart.

5.3 CART-POLE FLC

Before the issue of producing an adaptive GA-FLC can be addressed, a non-adaptive FLC must be designed to control the traditional cart-pole balancing system in which the mass of the cart remains constant. The procedure necessary to accomplish this task was presented in the previous chapter, and will be closely followed here.

The first step in developing the non-adaptive FLC is to decide on the controlled variables (these variables will be used to calculate errors and changes in error that appear on the left side of the rules which are of the form: IF <condition> THEN <action>). In the classic cart-pole balancing system, the four state variables listed in the previous section serve as the controlled variables. Thus, there will be two error terms: $E_x = s_x - x$ and $E_\theta = s_\theta - \theta$ where s_x and s_θ are the respective setpoints for cart position and pole angle ($s_x = s_\theta = 0.0$). There will also be two changes in error terms: $\Delta E_x = x$ and $\Delta E_\theta = \theta$. In a "common sense" approach, any decision on the action to be taken (the magnitude and direction of the force to be applied to the cart) must be based on the current value of these four variables. Next, a determination must be made as to what specific actions can be taken on the system, i.e., the manipulated variables must be determined. In the cart-pole balancing system, the only variable that is available for manipulation is the value of the force, F, to be applied to the center of mass of the cart.

The second step in the design of an FLC is the selection of linguistic terms to represent each of the controlled and manipulated variables. (As in the last chapter, the controlled variables do not appear in the rule set—E and ΔE appear instead.) Realize at the outset that there is not a definite method of doing this; the number and definition of the linguistic terms is always problem specific, and requires a general understanding of the system to be controlled. For this application, four linguistic terms were used to describe the error associated with the position of the cart, E_x. Three linguistic terms were deemed adequate to represent the variables, ΔE_x, E_θ, and ΔE_θ. The manipulated variable, F, necessitated the use of seven linguistic terms to be adequately represented. The specific linguistic terms used to describe the controlled and manipulated variables follow:

E_x Negative Big (**NB**), Negative Small (**NS**), Positive Small (**PS**) and Positive Big (**PB**);

ΔE_x Negative (**N**), Near Zero (**NZ**), and Positive (**P**);

E_θ Negative (**N**), Near Zero (**NZ**), and Positive (**P**);

ΔE_θ Negative (**N**), Near Zero (**NZ**), and Positive (**P**);

F Negative Big (**NB**), Negative Medium (**NM**), Negative Small (**NS**), Zero (**Z**),
Positive Small (**PS**), Positive Medium (**PM**) and Positive Big (**PB**).

After the linguistic terms have been chosen, they must be "defined." The linguistic terms are defined by membership functions. As with the initial requirement of selecting the necessary linguistic terms, there are no definite guidelines for constructing the membership functions; the terms are defined to represent the designer's general conception of what the terms mean. Membership functions can come in virtually any form. Two commonly used membership function forms (and the two forms that will be used in the applications presented in this chapter) are triangular and trapezoidal. The only restriction generally applied to the membership functions is that they have a maximum value of 1 and a minimum value of 0. When a membership function value is 1, there is complete confidence in the premise that the crisp value of the condition variable is accurately described by the particular linguistic term. When a membership function value is 0, there is complete confidence in the premise that the crisp value of the condition variable is ***not*** accurately described by the particular linguistic term. It is important to select membership functions that portray the developer's general conception of the linguistic terms. However, one should not anguish over the selection of the membership functions because the point of the GA-FLC link being presented is that help is on the way. A GA will be used to refine the membership functions to provide for near-optimal FLC performance. The membership functions developed by the authors for the cart-pole balancer appear in Figure 5.3.

The third step in the design of a FLC is the development of a rule set. The rule set in a FLC must include a rule for every possible combination of the variables as they are described by the chosen linguistic terms. Thus, 108 rules are required for the cart-pole balancing FLC as it has been designed to this point ($4 * 3 * 3 * 3 = 108$ possible combinations of the variables E_x, ΔE_x, E_θ, and ΔE_θ. Due to the nature of the linguistic terms, many of the actions needed for the 108 possible condition combinations are readily apparent. For instance, when the position of the cart is **NB**, the velocity of the cart is **N**, the position of the pole is **P**, and the angular velocity of the pole is **P**, the required action is without a doubt to apply a positive big force to the cart. However, there are some conditions for which the appropriate action is not readily apparent. In fact, there are some conditions for which the selection of an appropriate action seems almost contradictory. As an example, what is the appropriate action when the position of the cart is **PS**, the velocity of the cart is **P**, the position of the pole is **NZ**, and the angular velocity of the pole is **P**? The cart is to the right of the centerline and moving further away from the setpoint. Thus, if one is considering only the cart, the appropriate action would be to apply a small force in the negative direction. However, obviously one cannot consider only the cart. The state of the pole requires a small force in the positive direction. The traditional way to resolve these conflicts and to select an appropriate action has been to experiment with different selections of the action variables. As presented in the previous chapter, an alternative approach is to allow a GA to select an effective rule set for the membership functions as they have been defined by the developer. The rule set that is used for the cart-pole balancer was acquired the "old fashion way": trial-and-error directed by experience with controlling the system. The complete rule set used for the cart-pole balancing system appears in Figure 5.4. To help ensure that the reader understands this figure, the bold action in the figure is the appropriate action for the condition: E_x = **PS**, ΔE_x = **NZ**, E_θ = **P**, and ΔE_θ = **N**. Note that the linguistic terms for E_θ are read as **N**, **NZ**, and **P** from left to right, and the linguistic terms for ΔE_θ are read as **N**, **NZ**, and **P** from top to bottom. Thus, the complete rule is:

IF {E_x=**PS** AND ΔE_x=**NZ** AND E_θ=**P** AND ΔE_θ=**N**} THEN {F IS **NM**}.

Now that both the controller input variables and the manipulated variables have been chosen and described with linguistic terms, and a rule set has been written that prescribes an appropriate action for every possible set of conditions, it is left to determine a single crisp value of the force to be applied to the cart at a particular time step. This is a concern because more than one of the 108 possible rules can be applicable for a given state of the cart-pole system. A common technique for accomplishing this task is the Center of Area (COA) method [9] (sometimes called the centroid method). In the COA method, the action prescribed by each rule plays a part in the final crisp value of F. The contribution of each rule to the final value of F is proportional to the minimum confidence (the minimum value of the membership function values on the left side of the rule) one has in that rule for the specific state of the physical system at the particular time. This is equivalent to taking a weighted average of the prescribed actions. With the determination of a strategy for resolving "conflicts" in the actions prescribed by the individual rules, the FLC is complete.

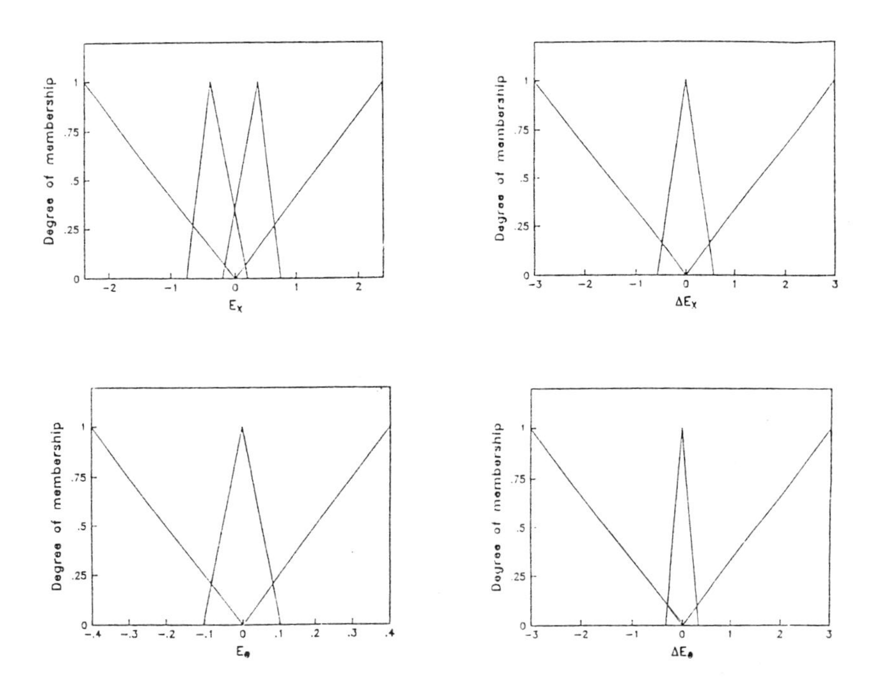

Figure 5.3: The author-developed membership functions used in the cart-pole balancing system for: [a] error in cart position (E_x), [b] change in error for cart position (ΔE_x), [c] error in pole position (E_θ), and [d] change in error for pole position (ΔE_θ).

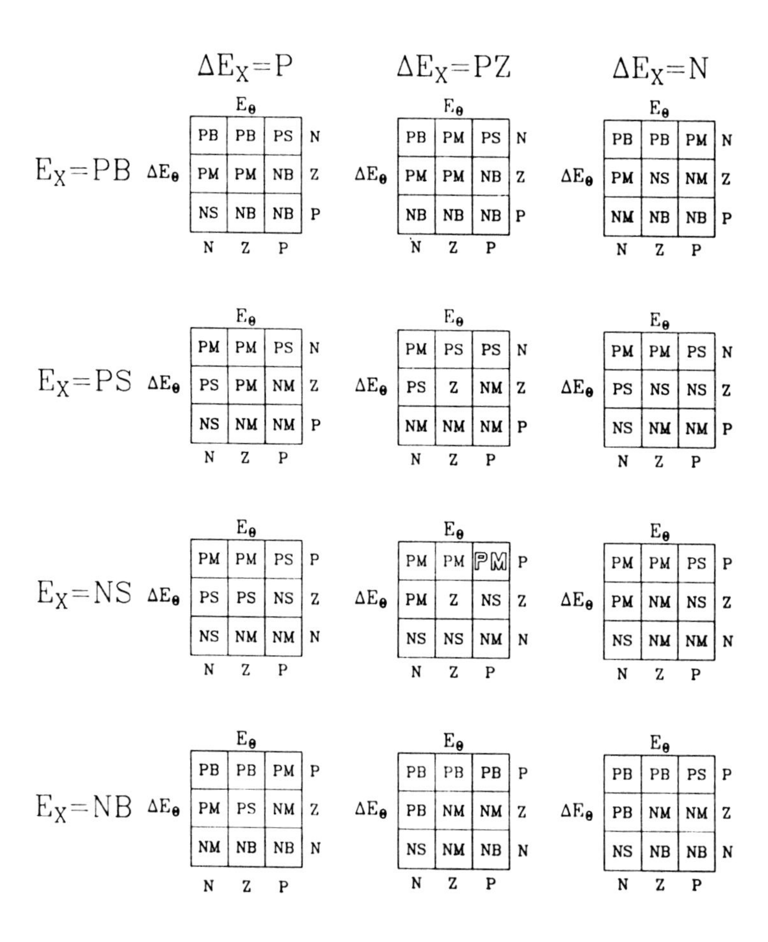

Figure 5.4: The complete rule set for the cart-pole balancing system includes 108 rules of the form: IF <condition> THEN <action>. Note that the linguistic terms for E_θ are read as **N**, **NZ**, and **P** from left to right, and linguistic terms for ΔE_θ are read as **N**, **NZ**, and **P** from top to bottom.

It is important to realize that to this point, the fact that the control problem is time-varying, i.e., that the mass of the cart changes, has not been broached. The time-varying aspect of the problem will be dealt with by altering the membership functions that define the linguistic terms describing the controlled and manipulated variables in real-time. This is akin to changing one's conception of what negative big actually is, for example, relative to the mass of the cart. Before the changing mass of the cart is considered, it seems logical to describe the way in which a GA is used to alter the membership functions to provide for near-optimal FLC performance.

As has been discussed in earlier chapters, there are numerous flavors of GAs; several genetic operators and variations of the basic scheme have been developed and implemented. In the previous chapter, one of these distinctive GAs, a micro GA, was introduced. The following discussion concerning the on-line selection of fuzzy membership functions is kept intentionally generic with respect to the particular GA employed. This is due to the fact that virtually any GA will provide better FLC performance, although in some problem domains one particular GA scheme may out-perform the others. For interested parties, the details of micro GAs (which is the scheme used in the research presented in this chapter) can be found in Karr [6]. Once the details of the particular GA to be employed have been determined, there are basically two decisions to be made when utilizing a GA to select FLC membership functions: (1) how to code the possible choices of membership functions as finite bit-strings and (2) how to evaluate the performance of the FLC composed of the chosen membership functions.

Consider the selection of a coding scheme. To define an entire set of triangular membership functions (functions for E_x, ΔE_x, E_θ, ΔE_θ, and F), several parameters must be selected. First, make the distinction between the two types of triangles used (see Figure 5.3). The right (90°) triangles appearing on the left and right boundaries will be termed ***extreme*** triangles, while the isosceles triangles appearing between the boundaries will be termed ***interior*** triangles. To completely define an extreme triangle, only one point must be specified because the apex of the triangle is fixed at the associated extreme value of the condition or action variable (the maximum value of **NB** for the error in cart position will always be at $E_x = -2.4$). On the other hand, the complete definition of an interior triangle necessitates the specification of two points, given the constraint that the triangles must be isosceles, i.e., the apex is at the midpoint of the two points specified. Thus, for the complete definition of a set of triangular membership functions for the cart-pole balancer as described above, 30 points had to be specified.

Certainly a 30 parameter search problem is challenging to say the least. Fortunately, the search space associated with the selection of membership functions for the cart-pole balancer can be pruned. Notice the rule set presented above is symmetric because every condition wherein the cart is to the left of the track's center has an analogous condition wherein the cart is to the right of the track's center. Therefore, **NB** should be "opposite" of **PB**, **NS** should be "opposite" of **PS**, and so on for all of the membership functions. Thus, instead of finding 30 parameters, the GA is faced with the task of finding only 15 points. Realize that even though the original search space has been reduced by a factor of two, a 15 parameter search problem is still of some consequence.

Now that the pertinent search parameters have been identified, a strategy for representing a set of these parameters as a finite bit-string must be developed. One such strategy that is popular, flexible, and effective is ***concatenated, mapped, unsigned binary coding***. In this coding scheme each individual parameter is discretized by mapping linearly from a minimum value (C_{min}) to a maximum value (C_{max}) using a 4-bit, unsigned binary integer according to the equation:

$$C = C_{min} + \frac{b}{(2^{l} - 1)} * (C_{max} - C_{min}) \quad (7)$$

where C is the value of the parameter of interest, and b is the decimal value represented by an l-bit string. Representing more than one parameter (such as the 15 parameters necessary in the cart-pole balancer) is accomplished simply by concatenating the individual 4-bit segments. Thus, in this example, a 60-bit string is necessary to represent an entire set of membership functions. This discretization of the problem produces a search space in which there exists $2^{60} = 1.152 * 10^{18}$ possible solutions.

Now what about the second decision? How are the strings, or the potential membership functions, evaluated? In judging the performance of the cart-pole FLC, it is important for the controller to center the cart and balance the pole. It should accomplish these tasks in the shortest time possible when initiated from any of a number of different initial conditions. These two objectives, centering the cart and balancing the pole, can be achieved by enticing the FLC to minimize a weighted sum of the absolute value of the distance between the cart and the center of the track and the absolute value of the difference between the angle of the pole and vertical, by altering the definition of the membership functions. The actual objective function the GA minimized in this study is:

$$f = \sum_{i=case1}^{i=case4} \sum_{j=0sec}^{j=30sec} (w_1 \mid x_{ij} \mid + w_2 \mid \theta_{ij} \mid) \qquad (8)$$

where $w_1 = 1.0$ and $w_2 = 10.0$ are weighting constants and the four cases are four different sets of initial conditions for the cart-pole system. Four initial condition cases were considered to ensure the FLC could accomplish the objective of centering the cart while balancing the pole from a variety of starting points.

To demonstrate the efficacy of using a GA to select membership functions, two sets of results are presented. First, results for the traditional cart-pole balancing system are given. Next, results are presented in which a GA is used to select membership functions on-line for the time-varying cart-pole balancing system.

Figure 5.5 shows results for the cart-pole balancer in which the mass of the cart remains constant. The author-developed FLC (AD-FLC) is compared to a FLC that uses membership functions selected by a micro GA (GA-FLC). The GA-FLC is able to achieve the goal of centering the cart and balancing the pole in approximately 7 seconds as compared to the 20 seconds required by the AD-FLC. Therefore, a GA has improved the initial design of the FLC. The question of whether or not this technique can be used to alter membership functions in real-time remains unanswered.

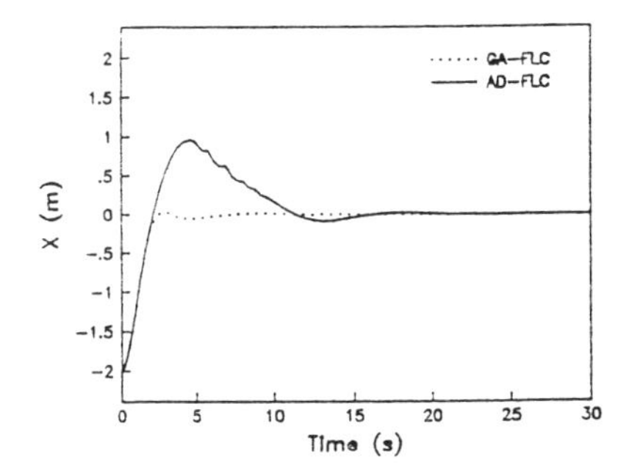

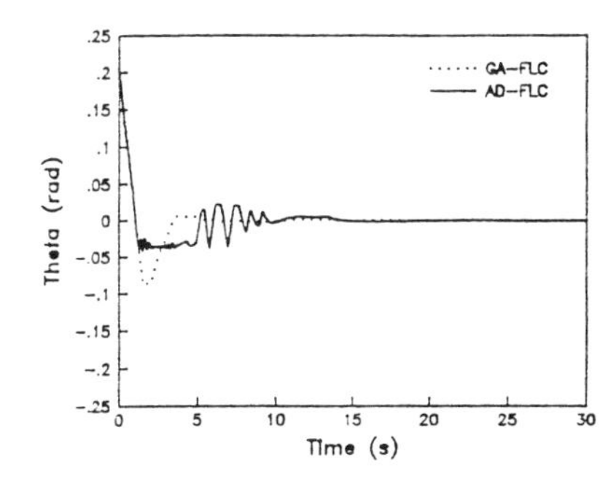

Figure 5.5: The GA-FLC outperforms the AD-FLC when considering both [a] cart position and [b] pole angle.

When the mass of the cart changes with time, the cart-pole balancing system responds differently to the application of forces. However, the mass of the cart was intentionally left out of the rule set. As it turns out, this was a prudent decision because changes in system response brought on by changes in cart mass can be adequately accounted for with alterations in the membership functions. The basic approach to using a GA to select high-performance functions as outlined in the preceding sections is not modified by the introduction of a time-varying parameter. However, some of the details needed to implement the technique are slightly different. For one, there is no need to alter the membership functions unless the mass of the cart changes. In the results presented, every time a change in mass occurs, a micro GA begins a search for new membership functions. Second, there is no need to look for robust membership functions that can accomplish the control objective from any set of initial conditions—what is needed is a set of membership functions that can accomplish the control objective beginning from the current state of the system. Therefore, the objective function does not have to include a summation over four initial condition cases, and the ***function evaluations*** associated with a GA are faster by a factor of four.

Figure 5.6 shows the results of an adaptive GA-FLC that accounted for changes in the cart mass. This adaptive GA-FLC was able to avoid the catastrophic failures of the pole falling over or the cart striking a wall despite the dramatic changes in cart mass (seen in Figure 5.2) by doing nothing more than altering its membership functions on-line. Every time a change in cart mass was made, a micro GA was employed to locate new membership functions that were effective for the current state of the system. As can be seen in Figure 5.6, the adaptive GA-FLC outperformed a non-adaptive FLC which allowed the pole to fall over.

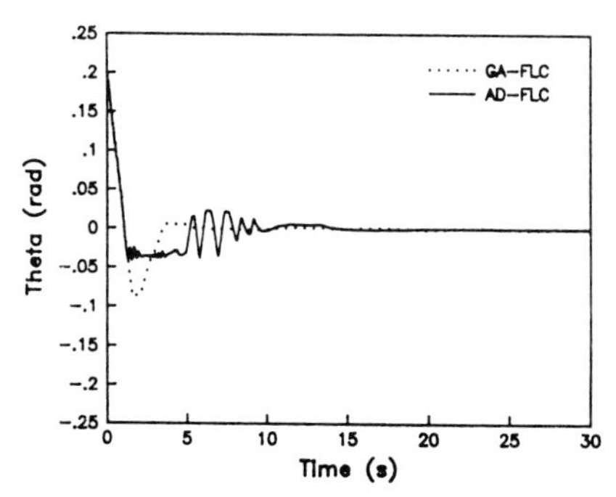

Figure 5.6: The adaptive GA-FLC was able to account for changes in cart mass to avoid catastrophic failure, unlike the non-adaptive FLC which allowed the pole to fall over.

The preceding has been an exposition on the use of a GA to develop an adaptive FLC for a cart-pole balancing system. The intent has been to provide sufficient detail for the reader to reproduce the results presented, or more realistically to at least grasp the concepts necessary for applying the technique of using GAs to improve FLC performance. The results presented above provide a glimpse of the power of the technique. In the next section, the robust nature of the technique is demonstrated when it is applied to a problem from the field of chemical engineering, namely the control of pH.

5.4 pH TITRATION SYSTEM

The power of the technique of applying GAs to the design of FLCs is further taxed when it is used to control a laboratory system (as opposed to a computer-simulated system). The extension to a laboratory system raises problems that simply do not exist in computer simulations, wherein there can be a tight reign over the environment being controlled. The laboratory pH system considered here is representative of the pH systems present in a number of mineral and chemical industries [8, 10]. It contains both nonlinearities and changing process dynamics. The nonlinearities are due to the fact that the output of pH sensors is proportional to the log of concentration, while the changing process dynamics are due to the addition of a buffer which significantly alters the manner that the pH responds as acid or base is added.

A schematic of the particular pH system considered, along with an adaptive architecture for controlling the system is shown in Figure 5.7. The system consists of a beaker initially containing a given volume of a solution having some known pH. There are five valved input streams into the beaker. The valves on only the two ***control input streams***, one a strong acid (0.1 M HCl) and one a strong base (0.1 M NaOH), can be adjusted by the controller. The other three valves are used to manipulate ***external streams*** which are altered by some "external agent," and thus a robust controller must be able to react to any changes in these input streams. Of these three streams, one is a strong acid (0.05 M HCl), one is a strong base (0.05 M CH_3COONa), and one is a buffer (a combination of 0.1 M CH_3COOH and 0.1 M CH_3COONa). The objective of the control problem is to neutralize the solution—drive the pH to 7—in the shortest time possible by adjusting the valves on the two control input streams. Additionally, the valves on the input streams are to be fully closed after the solution is neutralized. As a constraint on the control problem, the valves can only be adjusted a limited amount (0.5 mL/s/s, which is 20 pct of the maximum flow rate of 2.5 mL/s), thereby restricting the pressure transients in the associated pumping systems.

The pH system was designed on a small scale so that experiments could be performed in a laboratory of limited space. Titrations were performed in a 1,000-mL beaker with a magnetic stirring bar in the solution. Computer-driven peristaltic pumps were used for the five input streams. An industrial pH electrode and transmitter sent signals through an analog to digital board to a 33-MHz 386 personal computer which controlled the entire system.

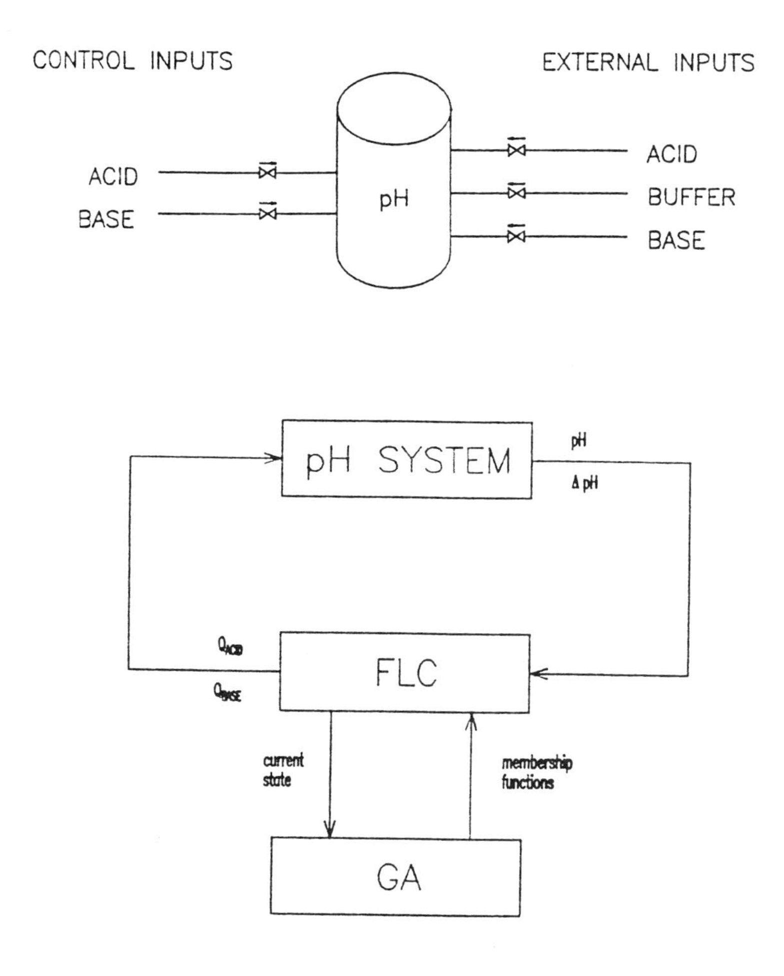

Figure 5.7: The laboratory system considered is difficult to control due to the nonlinearities present in pH, and due to the inclusion of external inputs.

To develop an adaptive FLC using GAs, a computer model of the physical system is required. This is due to the way in which the adaptive FLC is designed. While the FLC is actually manipulating the laboratory pH system, a GA is searching for improved membership functions by performing simulations of the pH system using a computer model. The GA's search is ideally performed in parallel with the FLC's manipulation of the laboratory system. However, due to limitations in computer hardware, this parallelism must often be simulated. Figure 5.7 shows a schematic of the basic design of an adaptive FLC that uses a GA for membership function selection. Fortunately, the dynamics of the pH system are well understood, and can be modelled using conventional techniques for buffered reactions [4]. The result is the following cubic equation that must be solved for $[H_3O^+]$ ions, which directly yields the pH of the solution:

$$x^3 + Ax^2 + Bx + C = 0 \tag{9}$$

where $x = [H_3O^+]$
$A = k_a + [CH_3COONa] + [NaOH] - [HCl]$,
$B = k_a[NaOH] - k_a[HCl] - k_a[CH_3OOH] - k_w$,
$C = -k_a k_w$,
$k_a = 1.8*10^{-5}$ is the equilibrium constant for CH_3COOH,
$k_w = 1.0*10^{-14}$ is the equilibrium constant for H_2O,

and bracketed terms [] represent molar concentrations. Further details of the computer model appear in a paper by Karr and Gentry [7]. In the pH system considered here, the development of a model of the physical system does not present an insurmountable obstacle. However, it should be realized that for many complex industrial systems, the development of an accurate computer model is, at the very least, an imposing task.

5.5 pH FLC

The laboratory pH system presented in the previous section can seem quite imposing due to the existence of external inputs, buffering agents, and the very nature of pH (logarithmic scale). However, the use of GAs for altering membership functions in real-time allow for the production of an adaptive FLC that is able to establish and maintain control of the laboratory system. In this section, the details of such a FLC are presented in much the same manner that the cart-pole FLC was presented in hopes that the repetition will reinforce the procedure.

Certainly there are numerous controlled variables that could be considered in the pH system (pH of solution in the tank, flowrates of the input streams, concentrations of input solutions, volume in the tank, and many others). However, recall that it is important to limit the number of controlled variables used because the size of the rule set increases multiplicatively with the number of controlled variables. After a period of experimentation (an inevitable requirement for the development of a quality FLC), two controlled variables were selected: the current value of pH (pH) in the beaker and the absolute value of the current time-rate-of-change of the pH in the tank (ΔpH). These controlled variables were used to describe an error (E=7-pH) and a change in error (ΔE=ΔpH). The fact that the pH system can be controlled when but two controlled variables are considered helps to demonstrate the degree to which the performance of a FLC depends on membership function values. As in the cart-pole system, the determination of the manipulated variables was relatively straightforward. There are basically only two things that can be altered by the controller: the valve settings (and thus the flowrates) associated with the control input streams. Therefore the two manipulated variables were the flowrates for the strong acid (Q_{ACID}) and the strong base (Q_{BASE}), respectively, of the input streams. The selection of the manipulated variables differs from the selection of the controlled variables in that the number of manipulated variables has no effect on the number of rules required by the FLC. Therefore, there is little point in limiting the number of manipulated variables considered.

Next, linguistic terms were selected to represent the controlled and manipulated variables. In both the liquid level example presented in the previous chapter, and in the cart-pole balancer example just discussed, the left hand side of all production rules included E and ΔE terms. Just so that the reader does not consider this a constraint, the left hand side of the rules in this pH example will use the controlled variables. Seven terms were used to describe pH, 2 terms were used to describe ΔpH, and 5 terms were used to describe both Q_{ACID} and Q_{BASE}. The specific linguistic terms used to describe the pertinent variables in the pH system follow:

pH — Very Acidic (**VA**), Acidic (**A**), Mildly Acidic (**MA**), Neutral (**N**), Mildly Basic (**MB**), Basic (**B**), and Very Basic (**VB**);

ΔpH — Small (**S**) and Large (**L**);

Q_{ACID} — Zero (**Z**), Very Small (**VS**), Small (**S**), Medium (**M**), and Large (**L**);

Q_{BASE} Zero (**Z**), Very Small (**VS**), Small (**S**), Medium (**M**), and Large (**L**).

Again, all of these linguistic terms are subjective, but the developers of the pH FLC (the authors) have some conception they associate with each of the terms.

The authors' conception of the linguistic terms is described by the membership functions which must be defined to give the terms some concrete meaning in the context of the FLC. The membership functions used to describe pH in the AD-FLC appear in Figure 5.8. The membership functions chosen for the pH FLC are trapezoidal (the membership functions used in the cart-pole FLC were triangular). These membership functions are later altered by a GA in response to changes occurring in the pH system. Alterations in these functions can dramatically change the performance characteristics of FLCs.

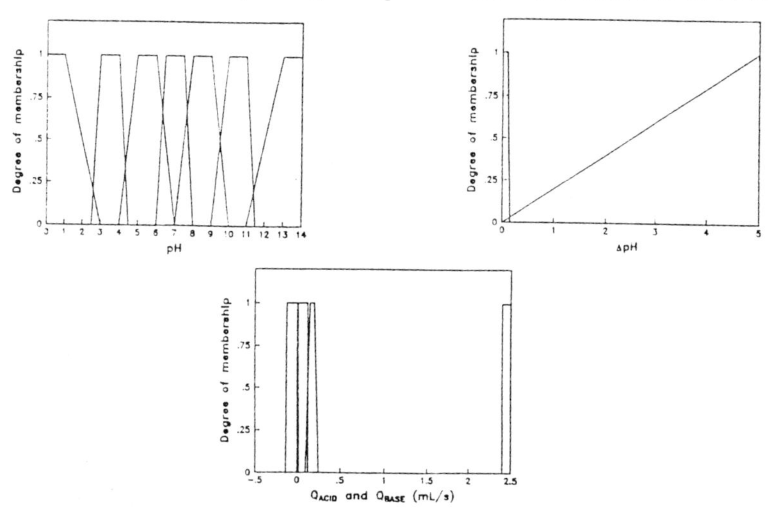

Figure 5.8: The AD membership functions used in the pH system for: [a] pH, [b] ΔpH, and [c] Q_{ACID} and Q_{BASE}.

Even though the laboratory pH system is arguably more complex than the cart-pole balancing system, an effective pH FLC can be written that contains far fewer rules than the cart-pole FLC. This is due, of course, to the fact that only two controlled variables are required in the pH system. The pH FLC required 2*7=14 rules to describe all of the possible conditions that could exist in the pH system as described by the linguistic terms represented by the membership functions shown in Figure 5.8. The entire rule set for the pH FLC is shown in Figure 5.9.

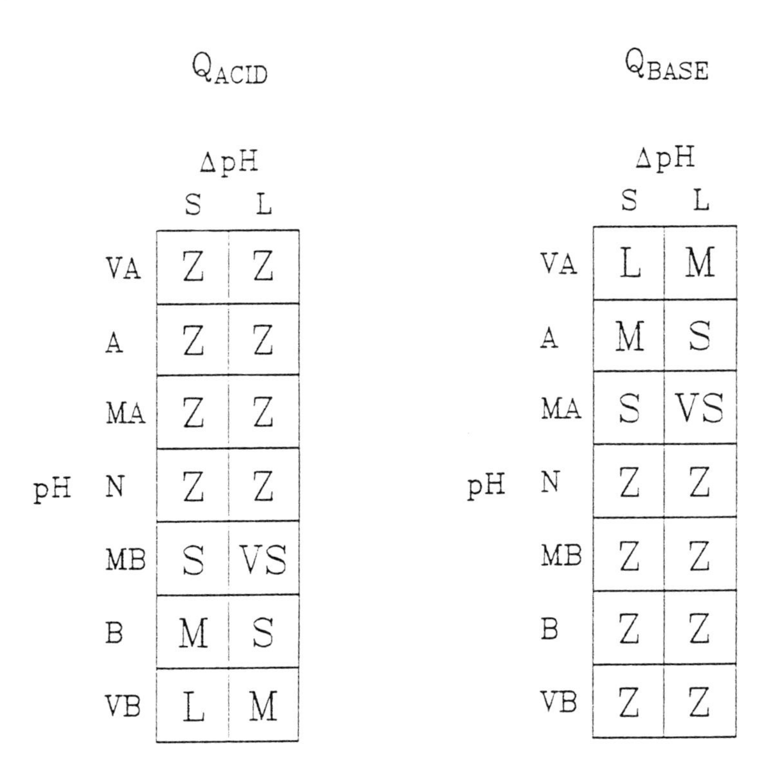

Figure 5.9: The complete rule set for the laboratory pH system includes 14 rules of the form: IF <condition> THEN <action>.

Now, the only aspect of the initial FLC design that is left to discuss is the technique for determining ***one*** value at which to set the flowrates of the input acid and base streams. As in the cart-pole balancing system, the popular COA method is used. One detail must be considered here. There is a limit on the allowable change in the flowrates of the input streams, i.e., the flowrates cannot change by more than 0.5 mL/s/s. However, the membership functions used in the COA method (shown in Figure 5.8) allow for values of Q_{ACID} and Q_{BASE} to range between 0.0 mL/s and 2.5 mL/s, irrespective of their current values. The constraint is imposed by computing the value of the flowrates using the COA method. If this value exceeds the constrained flowrate, the flowrate is changed by the maximum allowable value of 0.5 mL/s (for either increases or decreases in flowrate).

The preceding has been a general description of the elemental make-up of the pH FLC. At this point, the issue is with the use of a GA to alter the membership functions due to the changing process dynamics associated with the addition of a buffer. Furthermore, as with the cart-pole balancing system, these changes must be made in real-time. A GA is used to locate efficient membership functions in real-time, and the two basic decisions presented in the discussion of the cart-pole system must be made: (1) how to code the possible choices of membership functions as finite bit-strings and (2) how to evaluate the performance of the FLC composed of the chosen membership functions.

The coding scheme used is again concatenated, mapped, unsigned binary coding. In fact, the only difference between the coding used for the pH FLC and the cart-pole FLC is that more points are required to represent each individual membership function. Using the terminology introduced in the section on the cart-pole (***extreme*** and ***interior*** trapezoids), two points are required to define an extreme trapezoid while four points are required to define an interior trapezoid. This ideology is summarized in Figure 5.10.

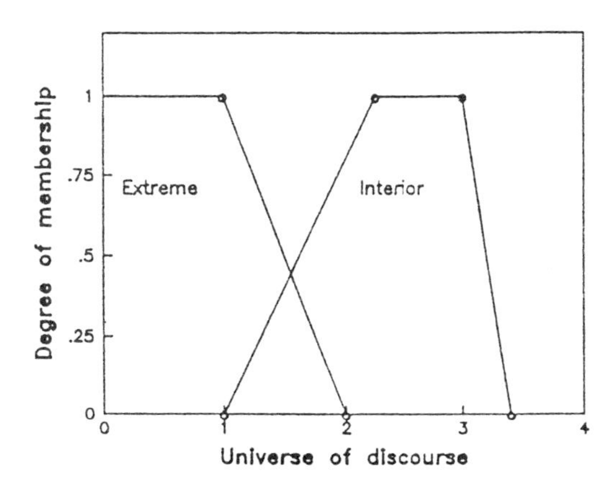

Figure 5.10: Four points are required to define each interior trapezoid while two points are required to define each extreme trapezoid.

If all of the membership functions described to this point were coded using concatenated, mapped, unsigned binary coding, the problem would be a 60 parameter search problem. Fortunately, two simplifications can be made which substantially reduce the size of the search space. First, the pH system is symmetric—everything above the setpoint of pH=7 has a mirror image below the setpoint. (However, this simplification cannot be made when considering a buffered system for symmetry about a pH of 7 is not present.) Therefore, the *functions* describing pH are symmetric. Second, the membership functions for Q_{ACID} and Q_{BASE} can be made identical. This is due to the fact that the concentrations of the input acid and input base are of the same strength. When these simplifications are made, the search space is reduced to 32 parameters. Seven bits were delegated to the representation of each parameter, thereby producing 224-bit strings, and a search space that includes $2^{224} = 2.696 * 10^{67}$ possible solutions.

The second decision that must be made is to determine a scheme for establishing the merit of each string. As with the cart-pole system, the objective of the pH FLC is to drive the pH in the tank to a setpoint (pH=7) in as short a time as possible, and to keep it there. Also, in the initial design phase, membership functions must be selected that are capable of accomplishing this control objective from any of a number of initial conditions. Therefore, the actual objective function that the GA minimized is:

$$f = \sum_{i=case1}^{i=case4} \sum_{j=0s}^{j=100s} (w_1 \mid 7.0 - pH \mid) + w_2(Q_{ACID} + Q_{BASE}) \tag{10}$$

When a GA is being used to select efficient membership functions in real-time, there is no need to consider each of the four initial condition cases. In fact, the only concern is to establish membership functions that are able to accomplish the control objective from the current state of the pH system. Therefore, the first summation in the above equation can be dropped, and the system simulation can be initiated from the current state of the laboratory system. This was the case in the on-line selection of membership functions for the cart-pole balancing system, and naturally reduces the length of time required by the GA to complete a function evaluation (determining the merit of one bit-string).

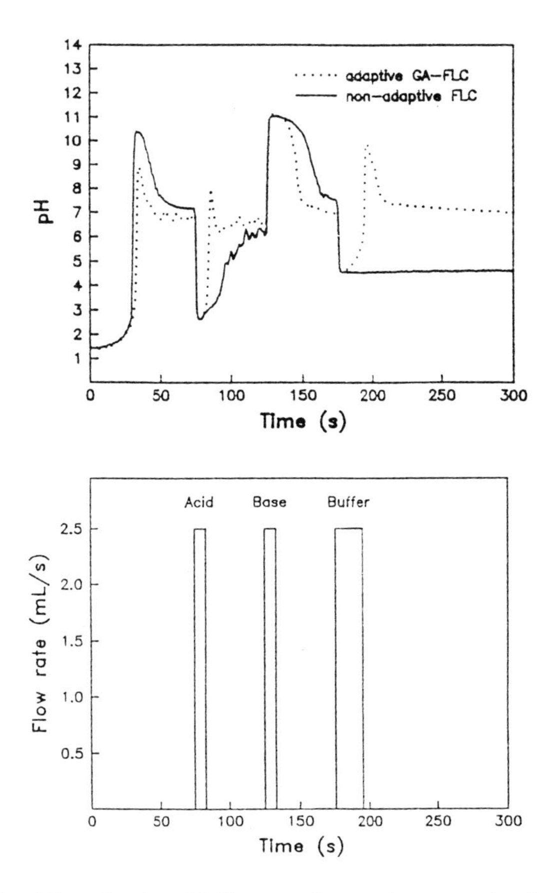

Figure 5.11: The adaptive FLC outperforms the non-adaptive FLC as shown in part [a] for the external inputs shown in [b].

The effect of using a micro GA to produce an adaptive pH-FLC is summarized in Figure 5.11. Figure 5.11a shows the flowrates of the three externally controlled input streams as a function of time. Figure 5.11b compares the performance of the non-adaptive AD-FLC with the performance of the adaptive GA-FLC. Notice that the adaptive GA-FLC was capable of accomplishing the control objective in approximately 75 s after the final external input which was a buffered solution. Although the non-adaptive AD-FLC was unable to accomplish the control objective in the 400 s allowed, it was eventually able to overcome the buffering effects and neutralize the solution. However, the additional time required by the non-adaptive FLC is substantial. It is also meaningful to note that both the non-adaptive and the adaptive FLCs were able to keep the pH of the solution in the tank under control despite "spikes" caused by the external addition of acid or base. This is because these additions represent perturbations to the system, not changes in the process dynamics like those caused by the addition of buffer solution.

5.6 GENO-FUZZY CONTROL OF BACKPROPAGATION

GAs are well suited for optimizing functions containing large number of variables as they are able to search the entire solution space simultaneously, offering fast convergence and avoiding being trapped in local minima.

The multilayer perceptron (MLP) network is used in this study. the exclusive-or function is chosen as a test case. There are two neurons in the input layer, four neurons in the hidden layer, and one neuron in the output layer. A GA has been used in the backpropagation learning algorithm to find the best set of weights for mapping the input data to the output data. Each string in the GA's population represents an entire set of weights. The weights are converted to a binary string using a concatenated mapped, unsigned binary coding. As a measure of the fitness of every string the mean squared error produced by the neural network when applying the weights contained in the string is considered. This error is a direct representation of the performance of the network with these weight values. The fitness of the string is calculated as:

$$fitness = 1.0 - MSE \tag{11}$$

where MSE = the mean squared error. The following parameters of GA are used in this implementation:

population size	=	80
chromosome length	=	192 bits
maximum range of weights	=	(-20, 20)
crossover probability	=	0.7
mutation probability	=	0.02

In the current implementation the weight optimization of the neural network is started by the GA, and as soon as a near-optimum configuration has been determined, the control is transferred to the fuzzy control for performing the final convergence of the weights to the optimum solution. The reason for this transfer is that the GA shows the very fast convergence in the first few iterations, but as it approaches the optimum solution, the speed of convergence decreases.

The fuzzy control is used to control the value of the learning rate, η, and the momentum term, α [5]. The change in the mean squared error (CE) between the target output and the actual output, and the change in CE (CEE) are used as crisp inputs to control η and α for fast convergence in the following equation:

$$\Delta W_{j\,k}(t + 1) = -(1.0 - \alpha)\ \eta\ \frac{\partial E}{\partial W_{j\,k}}(t + 1) + \alpha\ W_{j\,k}(t) \qquad (12)$$

where W_{jk} = weight from the jth neuron to the kth neuron. At every iteration the crisp values of CE and CEE are converted to fuzzy linguistic values using the fuzzification interface [13]. Tables 5.1 and 5.2 are used to control η and α. A commonly used defuzzification technique using the centroid method is employed to obtain a nonfuzzy control action.

CCE			**CE**		
	NL	**NS**	**ZE**	**PS**	**PL**
NL	NS	ZE	NS	NS	NS
NS	NS	PS	PS	ZE	NS
ZE	NS	ZE	PS	PS	NS
PS	NS	NS	PS	ZE	NS
PL	NS	NS	NS	NS	NS

Table 5.1: Fuzzy rules table used to control the learning parameter η.

CCE			CE		
	NL	NS	ZE	PS	PL
NL	NS	ZE	NS	NS	NS
NS	NS	PS	PS	ZE	NS
ZE	NS	ZE	PS	PS	NS
PS	NS	NS	PS	ZE	NS
PL	NS	NS	NS	NS	NS

Table 5.2: Fuzzy rules table used to control the momentum parameter α.

In this study a sigmoid activation function has been used in conjunction with a training tolerance of 0.2. The investigation shows that the geno-fuzzy control offers fast convergence and avoids getting stuck at local minima. The application of this approach to problems of larger scale are under investigation.

5.7 POTENTIAL AREAS OF RESEARCH

When this research was launched, the goal was to develop a comprehensive, stand-alone control system that could be used in the complex mineral processing systems used in industry today. Although the techniques presented in these two chapters have come a long way, they have not reached the required level of maturity. In this section, some research areas are presented that must be examined before the technology can reach its full potential.

Possibly the most challenging aspect of making this technology ready for industrial application, is to apply it to a physical system. This task has been accomplished with the application of the technology to a laboratory pH system. Given this fact, the following discussion of future research efforts will be focused on the pH system.

FLCs and GAs are, by their very nature, parallel algorithms; the implementation of the rules in FLCs is parallel while the evaluation of functions is parallel in GAs. Currently, the computer program that implements the adaptive pH FLC is simply simulating parallel processing through software. The use of transputers (parallel computing hardware) should allow for the implementation of the technology in hardware. When transputers are successfully employed, the entire control process will take less time. For one, the FLC will be able to determine an appropriate action in less time. Further, the GA will be able to locate high-performance membership functions more rapidly.

Once the system is made parallel, numerous doors of opportunity are opened. For instance, instead of discriminating between different approaches to making FLCs adaptive, numerous approaches can be implemented at the same time. In the previous chapter, two approaches were presented in which GAs could be used to improve FLC performance: (1) developing a rule set and (2) selecting membership functions. Based on previous experience, the authors are of the opinion that altering a FLC's rule set acts as a "course-tuning knob," while altering the membership functions can be thought of as a "fine-tuning knob." Changes in the rule set can cause dramatic changes in FLC performance. Sometimes the environment that is being controlled (for example the pH system) experiences such harsh changes that such a dramatic step as altering the rule set is necessary. As an example, if an extremely large quantity of buffer had been added to the pH system, the FLC may well have driven the system to its setpoint faster by altering its rule set instead of its membership functions. Other times (for example when a small amount of buffer is added to the pH system), changing the membership functions produces the most efficient controller. With parallel capabilities, different processing units can simultaneously investigate these two alternative approaches.

As they have been presented thus far, the GA-FLCs can be considered adaptive. Whether the extension to calling them "intelligent" systems can be made is another question all together. Certainly, there is much disagreement as to what constitutes an "intelligent" system, and that is part of the reason that the goal of this research has not been to develop an intelligent system. However, since these GA-FLCs are approaching some definitions of intelligent systems, consider the following discussion.

One requirement commonly associated with intelligent systems (aside from the one that they be adaptive) is that they possess some sort of memory and can distinguish between different situations. GA-FLCs could be provided with memory, but all of the details associated with accomplishing this task make it a less than trivial endeavor. As an example of a situation where a GA-FLC could use memory capabilities effectively, consider again the laboratory pH system. As it was presented in this chapter, a buffer was added one time and one time only. The addition of this buffer necessitated a change in the membership functions to efficiently reach the setpoint because the system dynamics had actually changed. If the system was acted upon in such a way as to return the system dynamics back to their original state (whether through the addition of a chemical to eliminate the buffer or if the buffer were physically removed from the tank), it would be expedient to simply reinstate those membership functions that had performed well before the buffer was added.

Another area of current interest is providing GA-FLCs with the ability to accomplish different objectives. For example, the goal in the pH system may not always be to drive the pH to a value of 7. Perhaps at some time an acidic solution is needed. It will be necessary to portray this need to the GA associated with the adaptive FLC. There are different ways to accomplish this communication, but it has not yet been done.

Producing an adaptive GA-FLC that is capable of changing its goals points to an even more ambitious task: to produce a comprehensive control system that is able to manage the everyday requirements of an industrial system. When such a system is actually developed for the systems used on a daily basis in the minerals processing industry, it will more than likely be a hybrid system; it will probably use techniques from the realms of fuzzy logic, genetic algorithms, neural networks, and traditional control. To develop such a hybrid system, decisions must be made as to when to employ the different techniques at the comprehensive controller's disposal. This type of comprehensive controller is the eventual goal of the research. Although headway has been made toward this goal, much work remains to be done.

5.8 SUMMARY AND CONCLUSIONS

This chapter was meant to be read in combination with the previous chapter. Chapter 4 strived to present a technique in which the search capabilities can be used to provide FLCs with adaptive capabilities. In Chapter 5, the goal has been to supply the reader with sufficient details so that this technique can actually be applied to his or her problem of interest.

The technique of combining GAs and FLCs has been applied to two systems: a simulated cart-pole balancing system and a laboratory pH system. In both instances, a micro GA was used to alter the membership functions used by a FLC. These alterations were made as a consequence of changes that occurred in the systems to be controlled, and were made in real-time. In both applications, the use of a GA to alter membership functions dramatically improved the performance of the FLCs. In the cart-pole balancing system, the GA allowed the FLC to actually avoid catastrophic failure. In the pH system, the GA simply allowed the FLC to drive the laboratory system to its setpoint in far less time, thus saving money.

The technique presented in these two chapters has not yet been developed to its fullest potential (Does any technology ever reach this point?). There are numerous aspects of the technology that must be investigated further. When these areas are investigated, the authors feel they will reveal further areas that must be examined. However, the results presented in this chapter indicate that using GAs to provide FLCs with adaptive capabilities is a powerful technique. The controllers produced with this technology are potentially beneficial in a number of industrial settings.

5.9 BIBLIOGRAPHY

[1] Anderson, C. W. (1987). "Strategy Learning with Multilayer Connectionist Representations", *Proceedings of the Fourth International Workshop on Machine Learning*, pp. 103-114.

[2] Barto, A. G., Sutton, R. S., and Anderson, C. W. (1983). "Neuronlike Adaptive Elements That Can Solve Difficult Learning Control Problems", *IEEE Transactions on Systems, Man, and Cybernetics*, vol. SMC-13, No. 5, pp. 834-846.

[3] Chen, Y. (1987). "Stability Analysis of Fuzzy Control — A Lyapunov Approach," *Proceedings of the IEEE International Conference on Systems, Man, and Cybernetics*, pp. 1027-1031.

[4] Hand, C. W., and Blewitt, G. L. (1986). *Acid-Base Chemistry*, New York, NY, USA, Macmillan Publishing Company.

[5] Haykin, S. (1994). *Neural Networks*, New York, NY, USA, Macmillan College Publishing Company.

[6] Karr, C. L. (1989). "Analysis and Optimization of an Air-Injected Hydrocyclone," Ph.D. Thesis, Tuscaloosa, AL, USA, The University of Alabama.

[7] Karr, C. L., and Gentry, E. J. (1992). "Real-Time pH Control Using Fuzzy Logic and Genetic Algorithms," *Proceedings of Annual Meeting of the Society for Mining, Metallurgy, and Exploration*, Phoenix, AZ.

[8] Kelly, E. G., and Spottiswood, D. J. (1982). *Introduction to Mineral Processing*, New York, NY, USA, John Wiley & Sons.

[9] Larkin, L. I. (1985). "A Fuzzy Logic Controller for Aircraft Flight Control", *Industrial Applications of Fuzzy Control* (M. Sugeno, ed.), Amsterdam, North-Holland, p. 87.

[10] McAvoy, T. J., Hsu, E., and Lowenthal, S. (1972). "Dynamics of pH in Controlled Stirred Tank Reactor", *Industrial Engineering Chemical Process Design and Development*, vol. 11, No. 1, pp. 68-70.

[11] Ross, T. J. (1995). *Fuzzy Logic With Engineering Applications*, New York, NY, USA, McGraw-Hill, Inc.

[12] Shah, I., and Rajamani, R. K. (1990). "A Self-Organizing Controller for Process pH Control", *Control '90 - Mineral and Metallurgical Processing* (R. K. Rajamani and J. A. Herbst, Eds.), p. 45.

[13] Vonk, E., et. at. (1995). "Integrating Evolutionary Computation with Neural Networks", ETD2000, IEEE Computer Society Press, USA, 137-143.

Chapter 6

Evolutionary engineering and applications

Hugo de GARIS
Brain Builder Group, Evolutionary Systems Department,
ATR Human Information Processing Research Laboratories,
2-2 Hikaridai, Seika-cho, Soraku-gun, Kyoto, 619-02, Japan
degaris@hip.atr.co.jp

This chapter reports on progress made in the first 3 years of ATR's "CAM-Brain" Project, which aims to use "evolutionary engineering" techniques to build/grow/evolve a RAM-and-cellular-automata based artificial brain consisting of thousands of interconnected neural network modules inside special hardware such as MIT's Cellular Automata Machine "CAM-8", or NTT's Content Addressable Memory System "CAM-CAM". The states of a billion (later a trillion) 3D cellular automata cells, and millions of cellular automata rules which govern their state changes, can be stored relatively cheaply in giga(tera)bytes of RAM. After 3 years work, the CA rules are almost ready. MIT's "CAM-8" (essentially a serial device) can update 200 million CA cells a second. It is likely that NTT's "CAM-CAM" (Cellular Automata on Content Addressable Memory) is essentially a massively parallel device, and will be able to update a *hundred billion* CA cells a second. Hence all the ingredients will soon be ready to create a revolutionary new technology which will allow thousands of evolved neural network modules to be assembled into artificial brains. This in turn will probably create not only a new research field, but hopefully a whole new industry, namely "brain building". Building artificial brains with a billion neurons is the aim of ATR's 8 year "CAM-Brain" research project, ending in 2001.

6.1 INTRODUCTION

ATR's CAM-Brain project resulted from the experience of the author's thesis work, in which he evolved neural net modules (using concatenated bit-string weights) to control the behavior of a simulated quadruped called "LIZZY", which could walk straight, turn left, turn right, peck at food and mate [6]. Each of these behaviors was controlled by the time varying outputs of a single evolved neural network module, and applied to the angles of the leg components of LIZZY. (As far as he is aware, the author was the first person to evolve neural net dynamics [3], (in the form of walking stick-legs "Walker")). Switching between behaviors involved taking the outputs from one neural net module and feeding them into the inputs of the next module. The next step was to evolve neural net detectors, e.g. for frequency, signal strength, signal strength difference, etc. Finally, neural net "production rule" modules were evolved which could map conditional inputs from detectors to output behaviors. Thus an "intelligent" artificial creature was built, which could detect prey, mates and predators, and then approach and eat or mate, or turn away and flee.

Virtually every neural net that the author tried to evolve, evolved successfully. *The evolution of these fully connected neural network modules proved to be a very powerful technique.* This success made a deep impression on the author, reinforcing his dream of being able to build much more complex artificial nervous systems, even artificial brains. However, every time the author added a neural net module to the Lizzy simulation, its speed on the screen was slowed (on a Mac II computer). Gradually, the necessity dawned on the author that some kind of evolvable hardware solution [5] would be needed to evolve large numbers of neural net modules and at great speed (i.e. electronic speed) in special machines the author calls "Darwin Machines" [5]. Evolving artificial brains directly in hardware remains the ultimate future goal of the author, but in the meantime (since the field of evolvable hardware (EHW, E-Hard) is today only in its infancy), the author compromises by using cellular automata to grow/evolve neural nets in large numbers in RAM, which is cheap and plentiful. (It is now possible to have a gigabyte (a billion byes) of RAM in one's work-station). By using cellular automata based neural nets which grow and evolve in gigabytes of RAM, it should be possible to evolve large numbers (tens of thousands) of neural net modules, and then assemble them (or even evolve their interconnections) to build an artificial brain. The bottleneck is the speed of the processor which updates the CA cells. State of the art in such processors is MIT's "CAM-8" machine, which can update 200,000,000 CA cells a second.

Recently, it has been suggested by the author's ATR colleague Hemmi, that NTT's Content Addressable Memory System "CAM-CAM" (which should be ready by the end of 1997) might be able to update CA cells at a rate *thousands* of times faster than the MIT machine, i.e. at a *hundred billion* CA cells per second. NTT's machine is massively parallel. Hemmi and his programmer assistant Yoshikawa are now (December 1995) busily engaged in writing software to convert the author's CA rules (in 2D form) into Boolean expressions suitable for the NTT machine. If they succeed in applying this machine to CAM-Brain, then a new era of brain building can begin, because the ability to evolve thousands of neural net modules would become realistic and very practical (for example, to evolve a neural net module inside a cubic space of a million CA cells, i.e. 100 cells on a side, at a hundred billion cells a second, would take at most about 500 clock cycles, i.e. about five milliseconds. *So the evolution of a population of 100 chromosomes over 100 generations could all be done in about one minute.)* All the essential ingredients for brain building would be available (lots of RAM, the CA rules, and fast CA processors). Even if Hemmi does not succeed, then a new machine can be designed to be thousands of times faster than the CAM-8 machine. The author believes the CAM-Brain breakthrough is either less than a year away, or at most only a few years away (the time necessary to design and build a "Super-CAM" machine, probably with the help of NTT).

The above gives an overview of the CAM-Brain research project. What now follows is a more detailed description of CAM-Brain, showing how one grows and evolves CA based neural net modules in 2D and 3D. We begin with the essential idea. Imagine a 2D CA trail which is 3 cells wide (e.g. Figure 6.2). Down the middle of the trail, send growth signals. When a growth signal hits the end of the trail, it makes the trail extend, or turn left, or right, or split etc., depending upon the nature of the signal (e.g. see Figures 6.3-6.6). It was the author who hand coded the CA rules which make these extensions, turns, splits etc. happen. The CA rules themselves are *not* evolved. It is the *sequence* of these signals (fed continuously over time into an initialized short trail) that is evolved. This sequence of growth signals is the "chromosome" of a genetic algorithm, and it is this sequence that maps to a cellular automata network. When trails collide, they can form "synapses" (e.g. see Figure 6.7). Once the CA network has been formed in the initial "growth phase", it is later used in a second "neural signaling phase". Neural signals move along CA-based axons and dendrites, and across synapses etc. The CA network is made to behave like a conventional artificial neural network (see Figure 6.11). The outputs of some of the neurons of the complex recurrent networks can be used to control complex time dependent behaviors whose fitnesses can be measured. These fitness values can be used to drive the evolution. By growing/evolving thousands of neural net modules and their interconnections in an incremental evolutionary way, it will be possible to build artificial brains. According to the CAM developers at MIT, it is likely that the next generation of CAMs will achieve an increase in performance of the order of thousands, within 5 years. However, to be able to evolve a billion neuron artificial brain by 2001 (ATR's goal), it is likely that a "nano-CAM" machine (i.e. one which uses nano-scale electronic speeds and densities) will need to be developed. To this end, we are collaborating with an NTT researcher who has developed a nanoscale electronics device, who wants to combine huge numbers of them to behave as nano-scale cellular automata machines.

In the summer of 1994, a two dimensional CAM-Brain simulation was completed which required 11,000 hand crafted CA state transition rules. It was successfully applied to the evolution of maximizing the number of synapses, outputting an arbitrary constant neural signal value, outputting a sine wave of a desired arbitrary period and amplitude and to the evolution of a simple artificial retina which could output the vector velocity of a "white line" which "moved" across an array of "detector" neurons. Work on the 3D simulation should be completed by early 1996, and is expected to take about 150,000 hand crafted CA rules. The Brain Builder Group of ATR took possession of one of MIT's CAM8 machines in the fall of 1994. At the time of writing (December 1995) the porting of the 2D rules from a Sparc20 workstation to the CAM8 is nearing completion. If the porting of the rules of the 3D simulation to this machine is not possible, then a "SuperCAM" machine will be designed specifically for CAM-Brain, with the collaboration of the Evolutionary Technologies (ET) group of NTT, with whom our Brain Builder group of ATR's Evolutionary Systems (ES) group, collaborates closely. The complexity of CAM-Brain will make it largely undesignable, so a (directed) evolutionary approach called "evolutionary engineering" is being used. Neural networks based on cellular automata (Codd 1968), can be grown and evolved at electronic speeds inside state of the art cellular automata machines, e.g. MIT's "CAM8" machine, which can update 200 million cells per second [10]. Since RAM is cheap, gigabytes of RAM can be used to store the states of the CA cells used to grow the neural networks. CA based neural net modules are evolved in a two phase process. Three cell wide CA trails are grown by sending a sequence of growth signals (extend, turn left, turn right, fork left, fork right, T fork) down the middle of the trail. When an instruction hits the end of the trail it executes its function. This sequence of growth instructions is treated as a chromosome in a Genetic Algorithm (Goldberg 1989) and is evolved. Once gigabytes of RAM and electronic evolutionary speeds can be used, genuine brain building, involving millions and later billions of artificial neurons, becomes realistic, and should become concrete within a year or two. The CAM-Brain Project should revolutionize the fields of neural networks and artificial life, and in time help create a new specialty called "Brain Building", with its own conferences and journals.

This chapter consists of the following sections. Section 2 describes briefly the idea of "Evolutionary Engineering", of which the CAM-Brain Project is an example. Section 3 describes how neural networks can be based on cellular automata (Codd 1968), and evolved at electronic speeds. Section 4 presents some of the details of CAM-Brain's implementation. Section 5 shows how using cellular automata machines will enable millions of artificial neural circuits to be evolved to form an artificial brain. Section 6 discusses changes needed for the 3D version of CAM-Brain. Section 7 deals with recent work. Section 8 deals with future work and section 9 summarizes.

6.2 EVOLUTIONARY ENGINEERING

Evolutionary Engineering is defined to be *"the art of using evolutionary algorithms (such as genetic algorithms (Goldberg 1989)) to build complex systems."* This chapter reports on the idea of evolving cellular automata based neural networks at electronic speeds inside cellular automata machines. This idea is a clear example of evolutionary engineering. Evolutionary engineering will be increasingly needed in the future as the number of components in systems grows to gargantuan levels. Today's nano-electronics for example, is researching single electron transistors (SETs) and quantum dots. Probably within a decade or so, humanity will have full blown nanotechnology (molecular scale engineering), which will produce systems with a trillion trillion components [8]. The potential complexities of such systems will be so huge, that designing them will become increasingly impossible. However, what is too complex to be humanly designable, might still be buildable, as this work will show. By using evolutionary techniques (i.e. evolutionary engineering), it is often still possible to *build* a complex system, even though one does not understand how it functions. This arises from the notion of the "complexity independence" of evolutionary algorithms, i.e. so long as the (scalar) fitness values which drive the evolution keep increasing, the internal complexity of the evolving system is irrelevant. This means that it is possible to successfully evolve systems which function as desired, but which are too complex to be designable. The author believes that this simple idea (i.e. the complexity independence of evolutionary algorithms) will form the basis of most 21st century technologies (dominated by nanotechnology [8]). Thus, evolutionary engineering can "extend the barrier of the buildable", but may not be good science, because its products tend to be black boxes. However, confronted with the complexity of trillion trillion component systems, evolutionary engineering may be the only viable method to build them.

6.3 CELLULAR AUTOMATA BASED NEURAL NETWORKS

Building an artificial brain containing billions of artificial neurons is probably too complex a task to be humanly designable. The author felt that brain building would be a suitable task for the application of evolutionary engineering techniques. The key ideas are the following. Use evolutionary techniques to evolve neural circuits in some electronic medium, so as to take advantage of electronic speeds. The medium chosen by the author was that of cellular automata (CA) (Codd 1968), using special machines, called "Cellular Automata Machines (CAMs)", which can update hundreds of millions of CA cells a second [10].

CAMs can be used to evolve the CA based neural networks at electronic speeds. The states of the cellular automata cells can be stored in RAM, which is cheap, so one can have gigabytes of RAM to store the states of billions of CA cells. This space is large enough to contain an artificial brain. MIT's Information Mechanics Group (Toffoli and Margolus) believe that within a few years it will be technically possible to update a trillion CA cells in about 0.1 nanoseconds (p221, [10]) Thus, if CA state transition rules can be found to make CA behave like neural networks, and if such CA based networks prove to be readily evolvable, then a potentially revolutionary new technology becomes possible. The CAM-Brain Project is based on the above ideas and fully intends to build artificial brains before the completion of the project in 2001. The potential is felt to be so great that it is likely that a new specialty will be formed, called "Brain Building".

For the first 18 months of the CAM-Brain Project, the author simulated a two dimensional version of CAM-Brain on a Sparc 10 workstation. This work was completed in the summer of 1994. The 2D version was used briefly (before work on the 3D version was started) to undertake some evolutionary tests, whose results will be presented in the next section. The 2D version served only as a feasibility and educational device. Since trails are obliged to collide in 2D, the 2D version was not taken very seriously. Work was begun rather quickly on the more interesting 3D version almost immediately after the 2D version was ready. Proper evolutionary tests will be undertaken once the 3D version is ready, which should be by early 1996. To begin to understand how cellular automata (Codd 1968) can be used as the basis for the growth and evolution of neural networks, consider Figure 6.1 which shows an example of a 2D CA state transition rule, and Figure 6.2 which shows a 2D CA trail, 3 cells wide. All cells in a CA system update the state of their cells synchronously. The new state of a given cell depends upon its present state and the states of its nearest neighbors. Down the middle of the 3 cell wide CA trail, move "signal or growth cells" as shown in Figure 6.2 As an example of a state transition rule which makes a signal cell move to the right one square, consider the right hand most signal cell in Figure 6.2, which has a state of 5. The cell immediately to its right has a state of 1, which we want to become a 5. Therefore the 2D state transition rule to turn the 1 into a 5 is 1.2.2.2.5→5. These signal or growth cells are used to generate the CA trails, by causing them to extend, turn left or right, split left or right, and Tsplit. When trails collide, they can form synapses. It is the sequence of these signal cells which determines the configuration of the CA trails, thus forming a CA network. It is these CA trails which later are used as neural network trails of axons and dendrites. Neural signals are sent down the middle of these CA trails. Thus there are two major phases in this process. Firstly, the CA trails are grown, using the sequence of signal cells. Secondly, the resulting CA trail network is used as a neural network, whose fitness at controlling some system can be measured and used to evolve the original growth sequence. To make this more explicit, it is the sequence of growth cells which is evolved. By modifying the sequence, one alters the CA network configuration, and hence the fitness of the configuration when it functions as a neural net in the second phase. From a genetic algorithm (GA) point of view, the format of the GA "chromosome" is the sequence of integers which code for the signaling or growth instructions. By mutating and crossing over these integers, one obtains new CA networks, and hence new neural networks. By performing this growth at electronic speeds in CAMs, and in parallel, with one CAM per GA chromosome, and attaching a conventional programmable microprocessor to each CAM to measure the user defined fitness of the CA based neural circuit, one has a means to evolve large numbers of neural modules very quickly. Using CAMs to evolve neural circuits, is an example of a type of machine that the author labels a "Darwin Machine", i.e. one which evolves its own structure or architecture. A related idea of the author concerns the concept of "Evolvable Hardware (EHW)" [5] where the software instructions used to configure programmable logic devices (PLDs) are treated as chromosomes in a Genetic Algorithm [9]. One then rewrites the circuit for each chromosome.

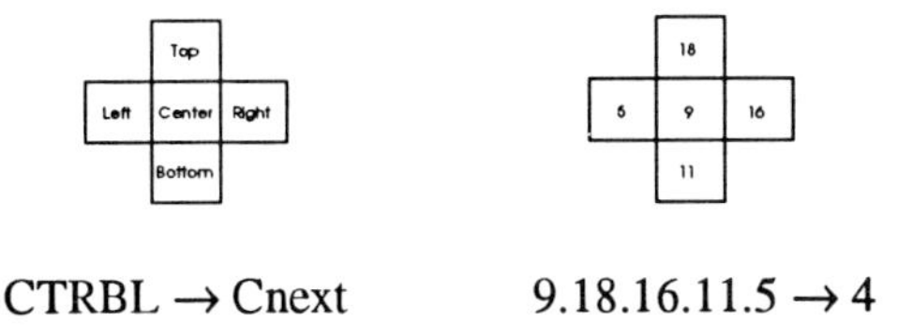

Figure 6.1: A 2D CA State Transition Rule

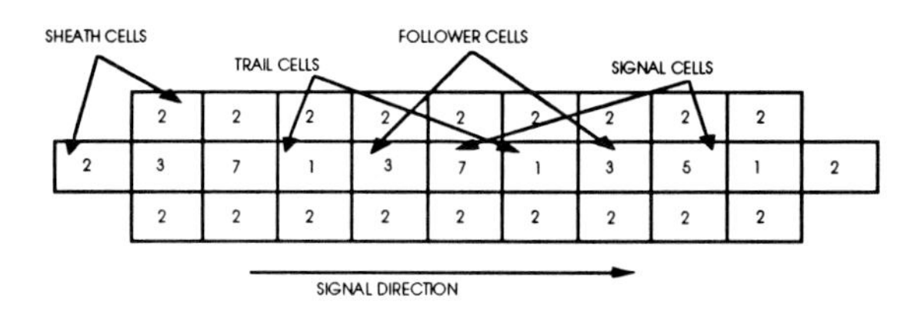

Figure 6.2: Signal Cells Move Along a Cellular Automata Trail

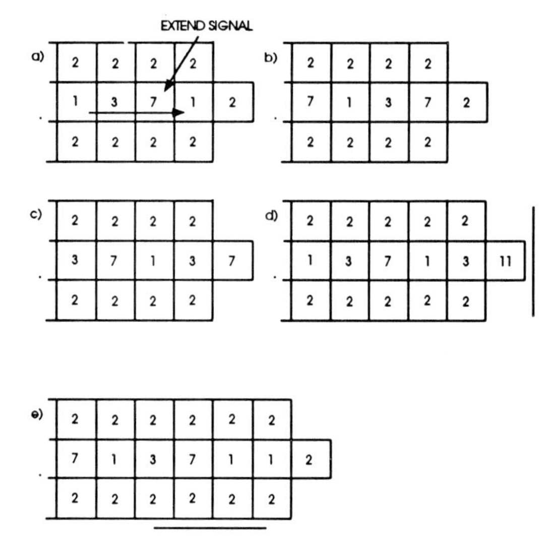

Figure 6.3: Extend the Trail

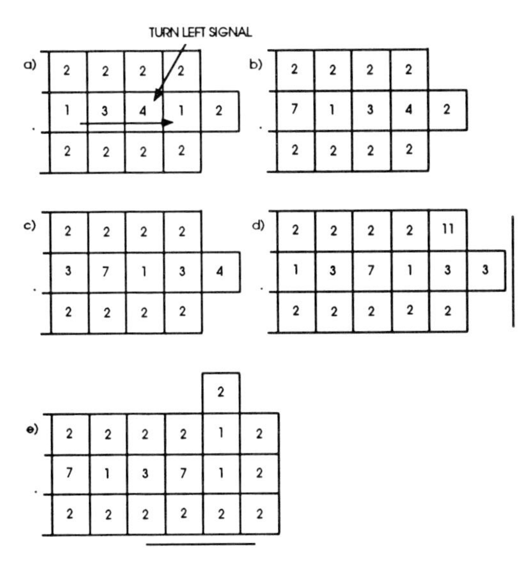

Figure 6.4 : Turn Trail Left

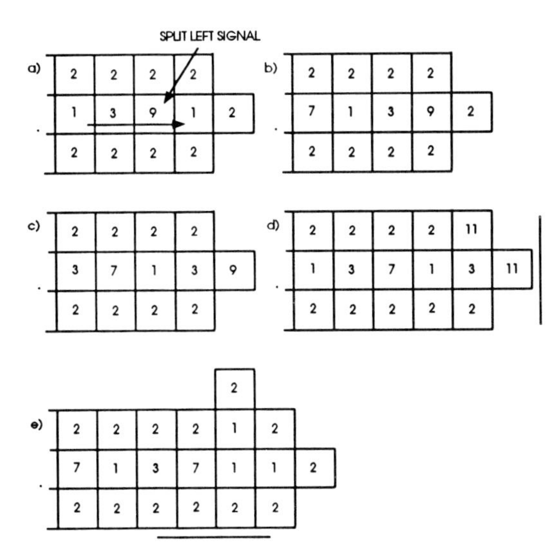

Figure 6.5 : Split Trail Left

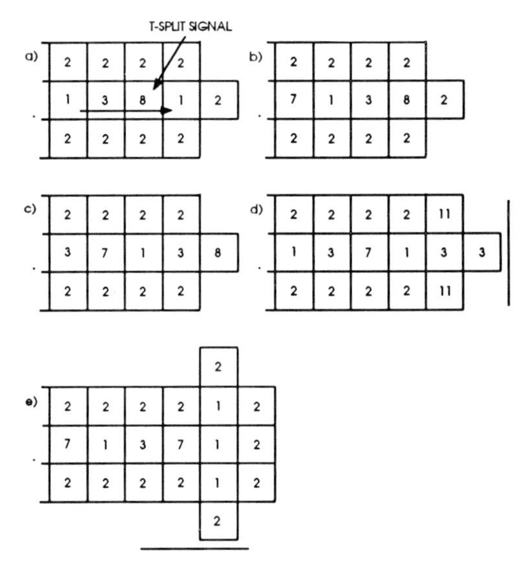

Figure 6.6: T-Split Trail

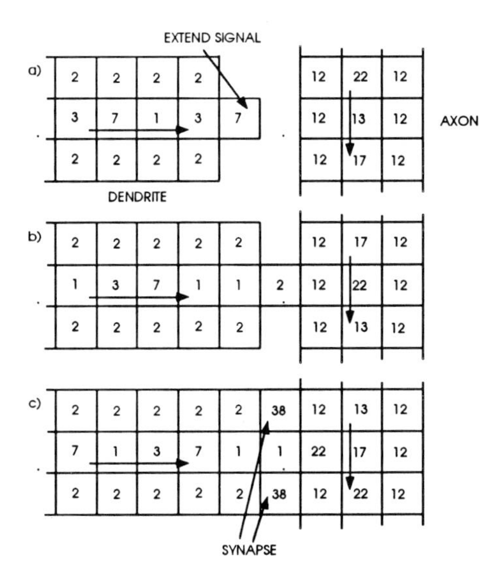

Figure 6.7 : Dendrite to Axon Synapsing

6.4 FURTHER DETAILS

This section provides further details on the implementation of the CA based neural networks. There are three kinds of CA trails in CAM-Brain, labeled dendrites, excitatory axons and inhibitory axons, each with their own states. Whenever an axon collides with a dendrite or vice versa, a "synapse" is formed. When a dendrite hits an excitatory/inhibitory axon or vice versa, an excitatory/inhibitory synapse is formed. An inhibitory synapse reverses the sign of the neural signal value passing through it. An excitatory synapse leaves the sign unchanged. Neural signal values range between -240 and +240 (or their equivalent CA states, ranging from 100 to 580). The value of a neural signal remains unchanged as it moves along an axon, but as soon as it crosses a synapse into a dendrite, the signal value (i.e. signal strength) begins to drop off linearly with the distance it has to travel to its receiving neuron. Hence the signal strength is proportional to the distance between the synapse and the receiving neuron. Thus the reduction in signal strength acts like a weighting of the signal by the time it reaches the neuron. But, this distance is evolvable, hence indirectly, the weighting is evolvable. CAM-Brain is therefore equivalent to a conventional artificial neural network, with its weighted sums of neural signal strengths. However, in CAM-Brain there are time delays, as signals flow through the network. When two or three dendrite signals collide, they sum their signal strengths (within saturated upper or lower bounds). When implementing the 2D version of CAM-Brain, it soon became noticeable that there were many many ways in which collisions between CA trails could occur. So many, that the author became increasingly discouraged. It looked as though it would take years of handcoding the CA state transition rules to get CAM-Brain to work. The intention was to have rules which would cover every collision possibility. Eventually a decision was made to impose constraints on the ways in which CA trails could grow. The first such constraint was to make the trails grow on a grid 6 cells or squares (cubes) on a side. This process (called "gridding") sharply reduces the number of collision types. It also has a number of positive side effects. One is that in the neural signaling phase, neural signals arrive synchronously at junction points. One no longer needs to have to handcode rules for phase delays in neural signaling summation. By further imposing that different growth cells advance the length of the trails by the same number of squares, one can further reduce the number of collision types. With synchrony of growth and synchrony of signaling and gridding, it is possible to cover all possible types of collisions. Nevertheless, it still took over 11000 rules to achieve this goal, and this was only for the 2D version. The 3D version is expected to take about 150,000 rules, but due to the experience gained in working on the 2D version, and to the creation of certain software productivity tools, the 3D version should be completed by early 1996.

Considering the fact that the 2D version takes 11,000 rules, it is impossible in this short chapter to discuss all the many tricks and strategies that are used to get CAM-Brain to work. That would require a book (something the author is thinking seriously about writing, if he ever makes time to do it). However, some of the tricks will be mentioned here. One is the frequent use of "gating cells", i.e. cells which indicate the direction that dendrite signals should turn at junctions to head towards the receiving neuron. To give these gating cells a directionality, e.g. a "leftness" or a "rightness", special marker cells are circulated at the last minute, after the circuit growth is stabilized. Since some trails are longer than others, a sequence of delay cells are sent through the network after the growth cells and before the marker cells. Without the delay cells, it is possible that the marker cells pass before synapses are formed. Once the 2D simulation was completed (before the CAM8 was delivered) several brief evolutionary experiments using the 2D version were undertaken. The first, was to see if it would be possible to evolve the number of synapses. Figures 6.9, 6.10, 6.11 show the results of an elite chromosome evolved to give a large number of synapses. Figure 6.9 shows early growth. Figure 6.10 shows completed growth, and Figure 6.11 shows the neural signaling phase. In this experiment, the number of synapses increased steadily. It evolved successfully. The next experiment was to use the neural signaling to see if an output signal (tapped from the output of one of the neurons) could evolve to give a desired constant value. This evolved perfectly. Next, was to evolve an oscillator of a given arbitrary frequency and amplitude, which did evolve, but slowly (it took a full day on a Sparc10 workstation). Finally, a simple retina was evolved which output the two component directional velocity of a moving "line" which passed (in various directions) over a grid of 16 "retinal neurons". This also evolved but even more slowly. The need for greater speed is obvious.

The above experiments are only the beginning. The author has already evolved (not using CAs) the weights of recurrent neural networks as controllers of an artificial nervous system for a simulated quadrupedal artificial creature. Neural modules called "GenNets" [2,3] were evolved to make the creature walk straight, turn left or right, peck at food, and mate. GenNets were also evolved to detect signal frequencies, to generate signal frequencies, to detect signal strengths, and signal strength differences. By using the output of the detector GenNets, it was possible to switch motion behaviors. Each behavior had its own separately evolved GenNet. By switching between a library of GenNets (i.e. their corresponding evolved weights) it was possible to get the artificial creature to behave in interesting ways. It could detect the presence and location of prey, predators and mates and take appropriate action, e.g. orientate, approach, and eat or mate, or turn away and flee. However, every time the author added another GenNet, the motion of the simulated creature slowed on the screen. The author's dream of being able to give a robot kitten some thousand different behaviors using GenNets, could not be realized on a standard monoprocessor workstation. Something more radical would be needed. Hence the motivation behind the CAM-Brain Project.

6.5 A BILLION NEURONS IN A TRILLION CELL CAM BY 2001

Fig. 6.8a shows some estimated evolution times for 10 chromosomes over 100 generations for a Sparc 10 workstation, a CAM8, and a CAM2001 (i.e. a CAM using the anticipated electronics of the year 2001) for a given application. In the 2D version of CAM-Brain, implemented on a Sun Sparc 10 workstation, it takes approximately 3.4 minutes to grow a stable cellular automata network consisting of only four neurons. It takes an additional 3.2 minutes to perform the signaling on the grown network, i.e. a total growth-signaling time to measure the fitness of a chromosome of 6.6 minutes. This time scales linearly with the number of artificial neurons in the network. If one uses a population of 10 chromosomes, for 100 generations, the total evolution time (on a Sparc 10) is 100*10*6.6 minutes, i.e., 110 hours, or 4.6 days. This is obviously tediously slow, hence the need to use a CAM. MIT's CAM8 [10] can update 25 million cellular automata cells per second, per hardware module. A CAM8 "box" (of personal computer size) contains eight such modules, and costs about $40,000. Such boxes can be connected blockwise indefinitely, with a linear increase in processing capacity. Assuming an eight module box, how quickly can the above evolution (i.e. 100 generations, with a population size of 10) be performed? With eight modules, 200 million cell updates per second is possible. If one assumes that the 2D CA space in which the evolution takes place is a square of 100 cells on a side, i.e., 10,000 cells, then all of these cells can be (sequentially) updated by the CAM8 box in 50 microseconds. Assuming 1000 CA clock cycles for the growth and signaling, it will take 50 milliseconds to grow and measure the fitness of one chromosome. With a population of 10, and 100 generations, total CAM8 evolution time for a four neuron network will be 50 seconds, i.e. about one minute, which is roughly 8000 times faster. Using the same CAM8 box, and a 3D space of a million cells, i.e. a cube of 100 cells on a side, one could place roughly 40 neurons. The evolution time will be 100 times as long with a single CAM8 box. With 10 boxes, each with a separate microprocessor attached, to measure the fitness of the evolved network, the evolution time would be about eight minutes. Thus for 1000 neurons, the evolution would take about 3.5 hours, quite an acceptable figure. For a million neurons, the evolution time would be nearly five months. This is still a workable figure. Note, of course, that these estimates are lower bounds. They do not include the necessary human thinking time, and the time needed for sequential, incremental evolution, etc. However, since the CAM-Brain research project will continue until the year 2001, we can anticipate an improvement in the speed and density of electronics over that period. Assuming a continuation of the historical doubling of electronic component density and speed every two years, then over the next eight years, there will be a 16-fold increase in speed and density. Thus the "CAM-2001" box will be able to update at a rate of 200*16*16 million cells per second. To evolve the million neurons above will take roughly 13.6 hours. Thus to evolve a billion neurons, will take about 19 months, again a workable figure. But, if a million neurons can be successfully evolved, it is likely that considerable interest will be focused upon the CAM-Brain approach, so that more and better machines will be devoted to the task, thus reducing the above 19-month figure. For example, with 100 machines, the figure would be about two months. The above estimates are summarized in Figure 6.8a. These estimates raise some tantalizing questions. For example, if it is possible to evolve the connections between a billion artificial neurons in a CAM2001, then what would one want to do with such an artificial nervous system (or artificial brain)? Even evolving a thousand neurons raises the same question.

Sparc10	CAM8	CAM8	CAM8	CAM8	CAM2001	CAM2001
10000 CA cells	10000 CA cells	1 million CA cells	25 million CA cells	25 billion CA cells	25 billion CA cells	25 trillion CA cells
4 neurons	4 neurons	40 neurons	1000 neurons	1 million neurons	1 million neurons	1 billion neurons
1 Sparc10	1 CAM8	10 CAM8s	10 CAM8s	10 CAM8s	10 CAM2001s	100 CAM2001s
4.6 days	50 seconds	8 minutes	3.5 hours	5 months	13.6 hours	2 months

Figure 6.8a: Evolution Times for Different Machines & CA Cell, Neuron & Machine Numbers

48*48*24	10gens 51	40gens 63	60gens 71	100gens 93	
96*48*24	10gens 81	20gens 89	45gens 122		
96*96*24	5gens 116	10gens 116	40gens 205	45gens 205	70gens 234
96*96*48	5gens 235	10gens 235			

Figure 6.8b: Synapses per Neuron Doubles as 3D Space Doubles

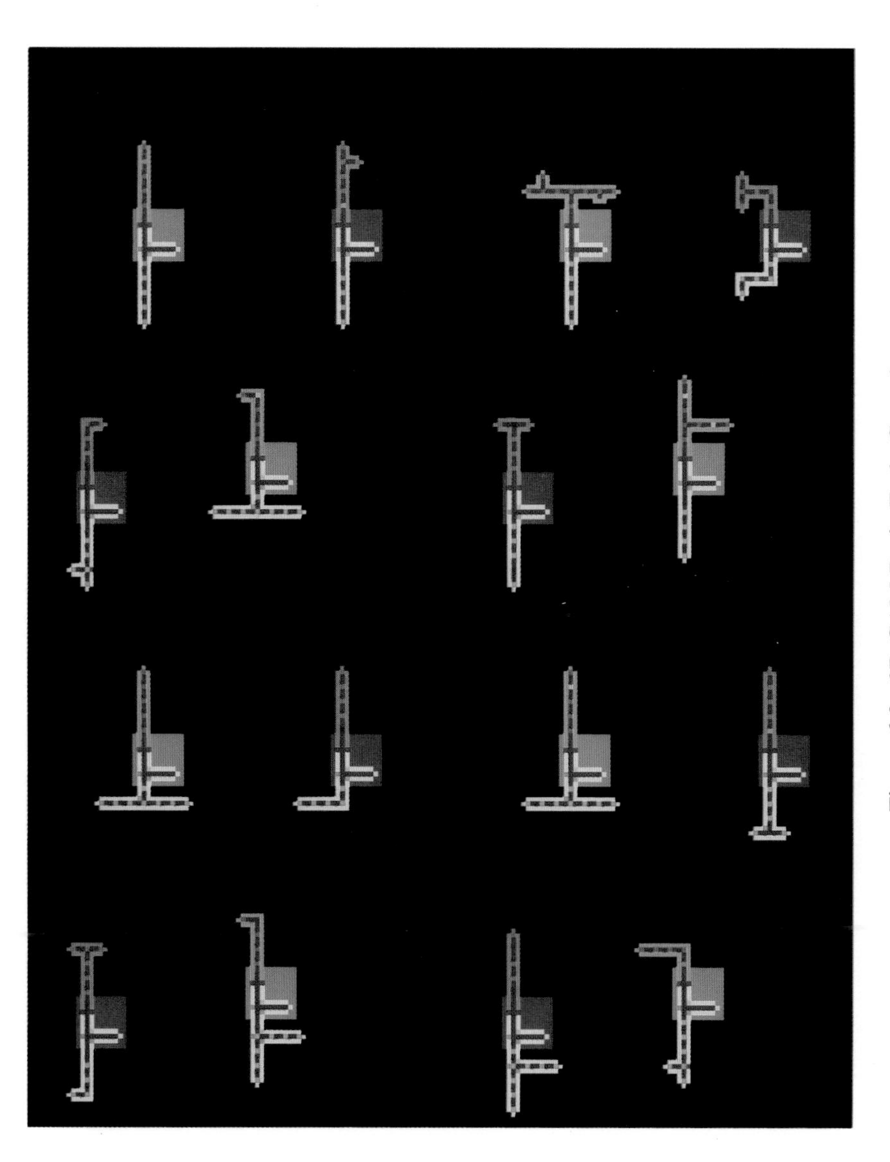

Figure 6.9: 2D CAM-Brain Early Growth

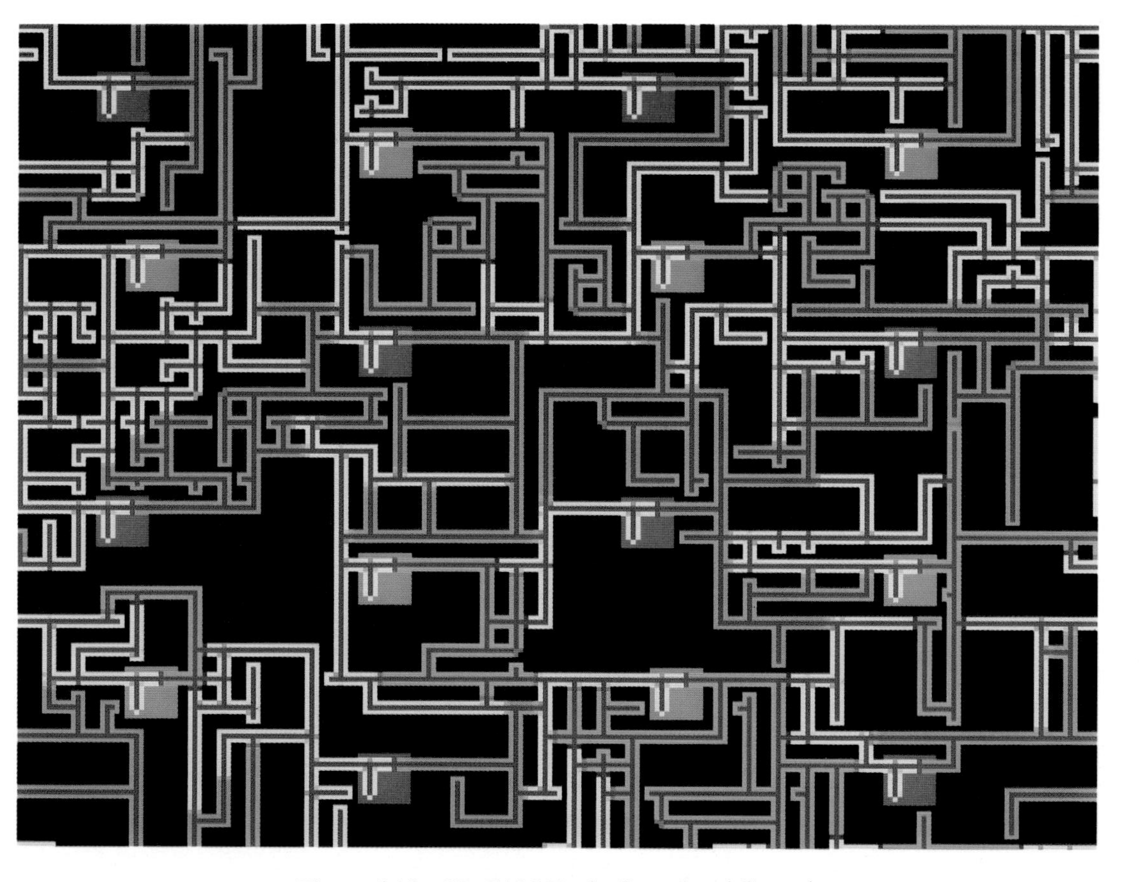

Figure 6.10: 2D CAM-Brain Completed Growth

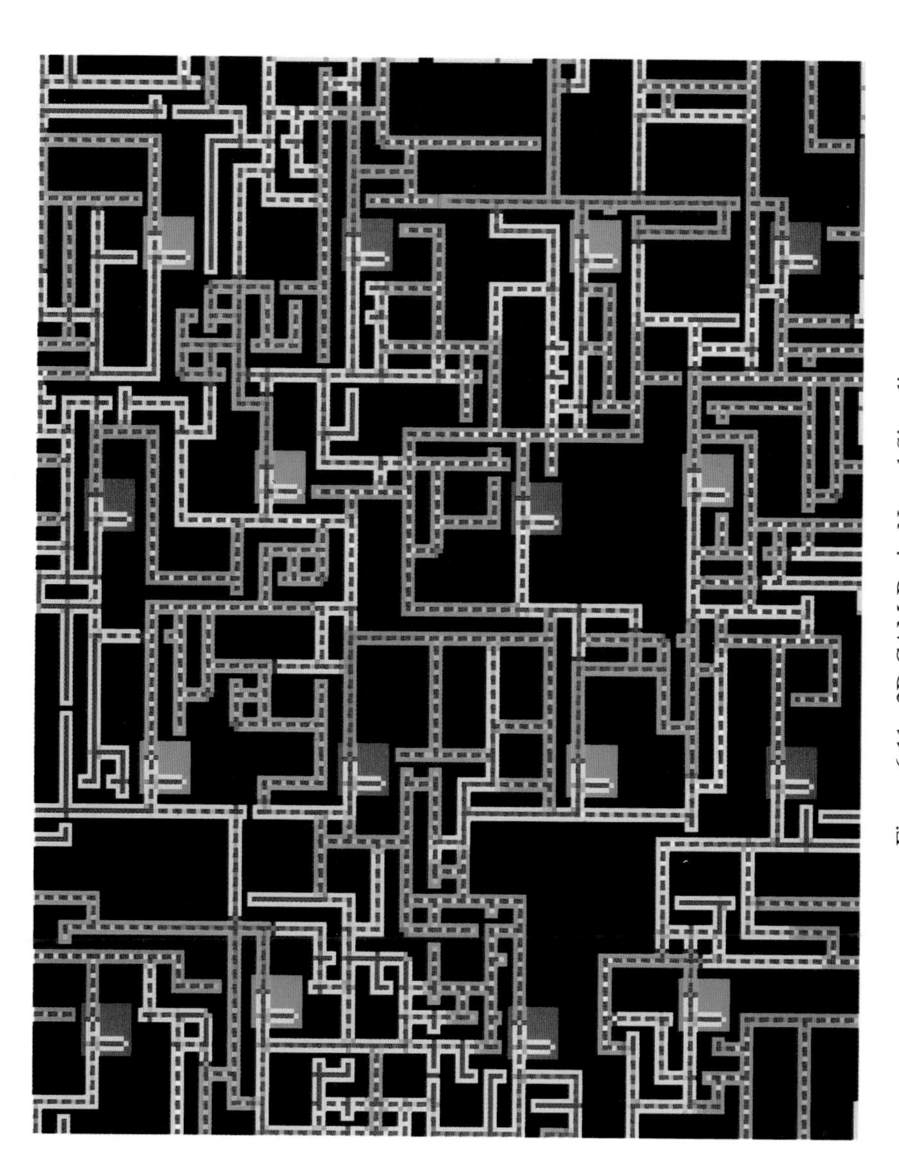

Figure 6.11: 2D CAM-Brain Neural Signaling

One of the aims of the CAM-Brain research project is to build an artificial brain which can control 1000 behaviors of a "robot kitten" (i.e. a robot of size and capacities comparable to a kitten) or to control a household "cleaner robot". Presumably it will not be practical to evolve all these behaviors at once. Most likely they will have to be evolved incrementally, i.e., starting off with a very basic behavioral repertoire and then adding (stepwise) new behaviors. In brain circuitry terms, this means that the new neural modules will have to connect up to already established neural circuits. In practice, one can imagine placing neural bodies (somas) external to the established nervous system and then evolving new axonal and dendral connections to it.

The CAM-Brain Project hopes to create a new tool to enable serious investigation of the new field of "incremental evolution." This field is still rather virgin territory at the time of writing. This incremental evolution could benefit from using embryological ideas. For example, single seeder cells can be positioned in the 3D CA space under evolutionary control. Using handcrafted CA "developmental or embryological" rules, these seeder cells can grow into neurons ready to emit dendrites and axons [4]. The CAM-Brain Project, if successful, should also have a major impact on both the field of neural networks and the electronics industry. The traditional preoccupation of most research papers on neural networks is on analysis, but the complexities of CAM-Brain neural circuits, will make such analysis impractical. However, using Evolutionary Engineering, one can at least build/evolve functional systems. The electronics industry will be given a new paradigm, i.e. evolving/growing circuits, rather than designing them. The long term impact of this idea should be significant, both conceptually and financially.

6.6 3D VERSION

The 3D version is a conceptually (but not practically) simple extension of the 2D version. Instead of 4 neighbors, there are 6 (i.e. North, East, West, South, Top, Bottom). Instead of 6 growth instructions as in the 2D version (i.e. extend, turn left, turn right, split extend left, split extend right, split left right), there are 15 in the 3D version. A 3D CA trail cross section consists of a center cell and 4 neighbor cells, each of different state or color (e.g. red, green, blue, brown). Instead of a turn left instruction being used as in the 2D case, a "turn green" instruction is used in the 3D case. The 15 3D growth instructions are (extend, turn red, turn green, turn blue, turn brown, split extend red, split extend green, split extend blue, split extend brown, split red brown, split red blue, split red green, split brown blue, split brown green, split blue green). A 3D CA rule thus consists of 8 integers of the form CTSENWB→Cnew. The 3D version enables dendrites and axons to grow past each other, and hence reach greater distances. The weakness with the 2D version is that collisions in a plane are inevitable, which causes a crowding effect, whereby an axon or dendrite cannot escape from its local environment. This is not the case with the 3D version, which is topologically quite different. A 3D version is essential if one wants to build artificial brains with many interconnected neural modules. The interconnectivity requires long axons/dendrites. Figure 6.12 shows an early result in 3D simulation. A space of 3D CA cells (48*48*48 cubes) was used. A single short 3D CA trail was allowed to grow to saturate the space. One can already sense the potential complexity of the neural circuits that CAM-Brain will be able to build. In 3D, it is likely that each neuron will have hundreds, maybe thousands of synapses, thus making the circuits highly evolvable due to their smooth fitness landscapes (i.e. if you cut one synapse, the effect is minimal when there are hundreds of them per neuron).

6.7 RECENT WORK

Just prior to writing this chapter, the author was able to test the idea that in 3D a single neuron could have an arbitrarily large number of synapses, provided that there is enough space for them to grow in. This was a crucial test, whose results are shown in Figure 6.8b. Fitness was defined as the number of synapses formed for two neurons in CA spaces of 48*48*24, 96*48*24, 96*96*24, and 96*96*48 cells respectively. One can see that by doubling the space, one doubles (roughly) the number of synapses (for the elite chromosome). If this had not been the case, for example, if some kind of fractal effect had caused a crowding of the 3D circuits (similar to the crowding effect in 2D), then the whole CAM-Brain project would have been made doubtful. However, with this result, it looks as though evolvability in the 3D signaling phase will be excellent, although the author needs several months more work before completing the 3D signaling phase to confirm his confidence.

At the time of writing (December 1995), the author is completing the simulation of the 3D version, working on the many thousands of rules necessary to specify the creation of synapses. So far, more than 140,000 3D rules have been implemented, and it is quite probable that the figure may go higher than 150,000. Since each rule is rotated 24 ways (6 ways to place a cube on a surface, then 4 ways to rotate that cube) to cater to all possible orientations of a 3D trail, the actual number of rules placed in the (hashed) rule base will be over 3 million. Specifying these rules takes time, and constitutes so far, the bulk of the effort

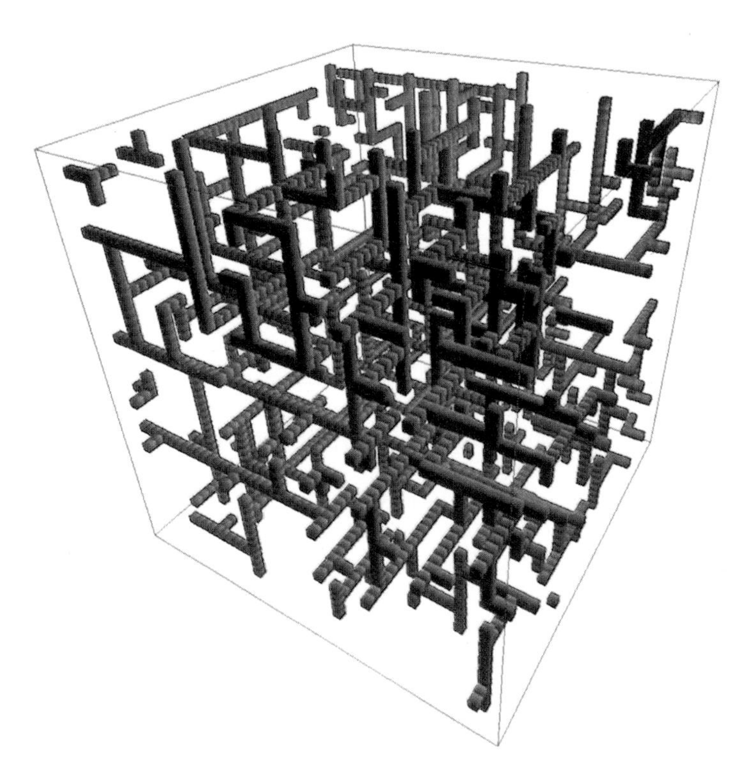

Figure 6.12: 3D CAM-Brain Non-Synaptic Growth

spent building the CAM-Brain system. Software has been written to help automate this rule generation process, but it remains a very time consuming business. Hence the immediate future work will be to complete the simulation of the 3D version. Probably, this will be done by early 1996.

Early in 1995, the author put his first application on the CAM8 machine (which rests on his desk). MIT's CAM8 is basically a hardware version of a look up table, where the output is a 16 bit word which becomes an address in the look up table at the next clock cycle. This one clock cycle lookup loop is the reason for CAM8's speed. It is possible to give each CA cell in the CAM8 more than 16 bits, but tricks are necessary. The first CAM8 experiment the author undertook involved only 16 bits per CA cell. This chapter is too short to go into details as to how the CAM8 functions, so only a broad overview will be given here. The 16 bits can be divided into slices, one slice per neighbor cell. These slices can then be "shifted" (by adding a displacement pointer) by arbitrarily large amounts (thus CAM8 CA cells are not restricted to having local neighbors). With only 16 bits, and 4 neighbors in the 2D case (Top, Right, Bottom, Left) and the Center cell, that's only 3 bits per cell (i.e. 8 states, i.e. 8 colors on the display screen). It is not possible to implement CAM-Brain with only 3 bits per CA cell. It was the intention of the author to use the CAM8 to show its potential to evolve neural circuits with a huge number of artificial neurons. The author chose an initial state in the form of a square CA trail with 4 extended edges. As the signals loop around the square, they duplicate at the corners. Thus the infinite looping of 3 kinds of growth signals supply an infinite number of growth signals to a growing CA network. There are 3 growth signals (extend, extend and split left, extend and split right). The structure needs exactly 8 states. The 8 state network grows into the 32 megacells of 16 bits each, which are available in the CAM8. At one pixel per cell, this 2D space takes over 4 square meters of paper poster (hanging on the author's wall). A single artificial neuron can be put into the space of one's little finger nail, thus allowing 25,000 neurons to fit into the space. If 16 Mbit memory chips are used instead of 4 Mbit chips, then the area and the number of neurons quadruples to 100,000.

Placing the poster on the author's wall suddenly gave visitors a sense of what is to come. They could see that soon a methodology will be ready which will allow the growth and evolution of artificial brains, because soon it will be possible to evolve many thousands of neural modules and their inter-connections. The visitors sense the excitement of CAM-Brain's potential.

Filling a space of 32 Mcells, with artificial neurons can be undertaken in at least two ways. One is to use a very large initialization table with position vectors and states. Another, is to allow the neurons to "grow" within the space. The author chose to use this "neuro-embryonic" approach. A single "seeder" CA cell is placed in the space. This seeder cell launches a cell to its right and beneath it. These two launched cells then move in their respective directions, cycling through a few dozen states When the cycle is complete, they deposit a cell which grows into the original artificial neuron shape that the author uses in the 2D version of CAM-Brain. Meanwhile other cells are launched to continue the growth. Thus the 32Mcell space can be filled with artificial neurons ready to accept growth cell "chromosomes" to grow the neural circuitry. This neuro-embryogenetic program (called "CAM-Bryo") was implemented on a workstation by the author, and ported to the CAM8 by his research colleague Felix Gers. In order to achieve the porting, use was made of "subcells" in the CAM8, a trick which allows more than 16 bits per CA cell, but for N subcells of 16 bits, the total CAM8 memory space available for CA states is reduced by a factor of N. Gers used two subcells for CAM-Bryo, hence 16M cells of 32 bits each. A second poster of roughly two square meters was made, which contained about 25,000 artificial neurons (see Figure 6.13). Again, with 16Mbit memory chips, this figure would be 100,000. Gers expects to be able to port the 2D version of CAM-Brain to the CAM8 with a few weeks work, in which case, a third poster will be made which will depict about 15,000 neurons (with a lower density, to provide enough space for the neural circuitry to grow) and a mass of complex neural circuits. Once this is accomplished, we expect that the world will sit up and take notice - more on this in the next section.

The author's boss at ATR's Evolutionary Systems department, has recently set up a similar group at his company NTT, called Evolutionary Technologies (ET) department. The idea is that once the ATR Brain Builder group's research principles are fairly solid, the author and the author's boss (whose careers are now closely linked) will be able to tap into the great research and development resources of one of the world's biggest companies, when the time comes to build large scale artificial brains. NTT has literally thousands of researchers.

The author would like to see Japan invest in a major national research project within the next 10 years to build "Japan's Artificial Brain", the so-called "J-Brain Project". This is the goal of the author, and then to see such a project develop into a major industry within 20 years. Every household would like to have a cleaner robot controlled by an artificial brain. The potential market is huge.

6.8 FUTURE WORK

A lot of work remains to be done. The author has a list of "to dos" attached to his computer screen. The first item on the list is of course, to finish the rules for the

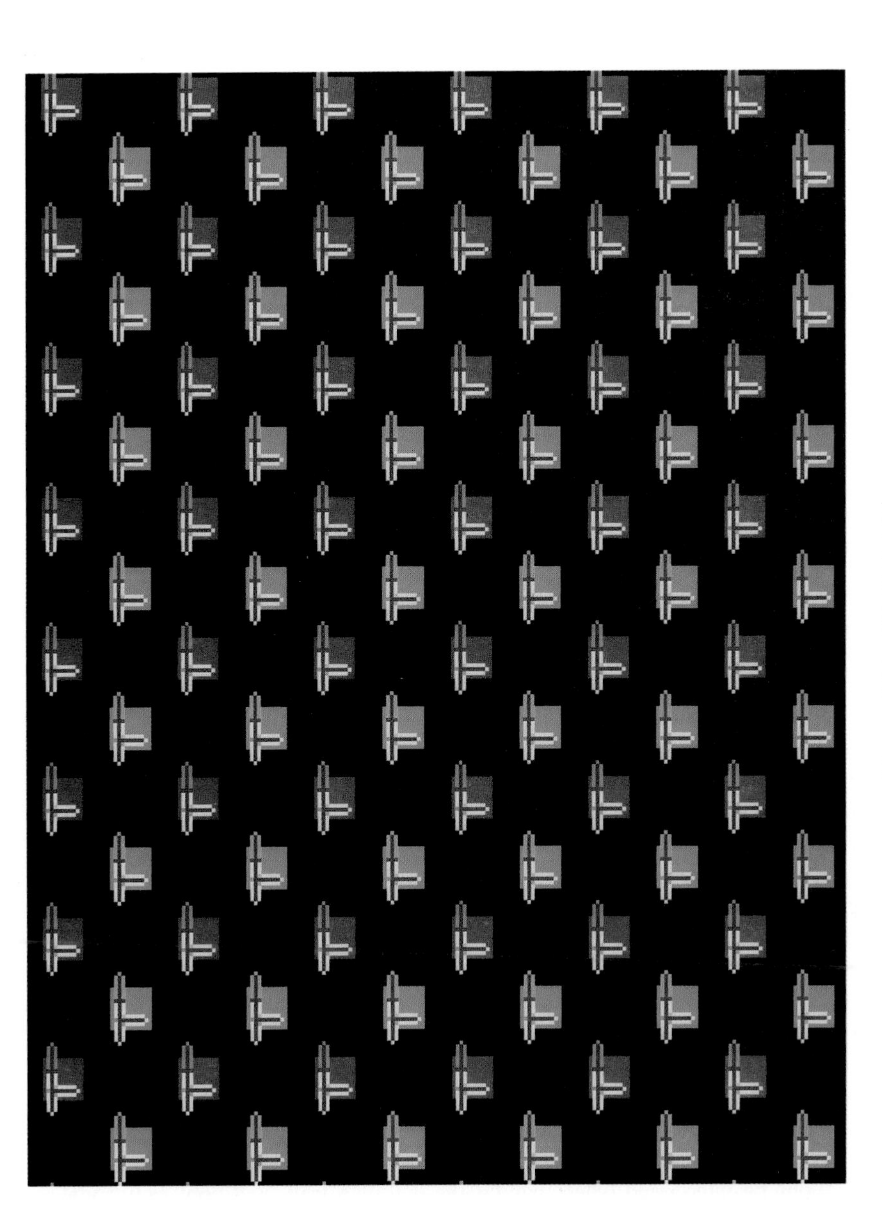

Figure 6.13: 2D CAM-Bryo

3D version of CAM-Brain. This should be done by early 1996, and will probably need over 150,000 CA rules. Second, the experience gained in porting the 700 rules for "CAM-Bryo" from a workstation to the CAM8 will shortly enable Gers to complete the much tougher task of porting the 2D version of CAM-Brain to the CAM8. In theory, since there are 11,000 CA rules for the 2D version, and that each rule has 4 symmetry rotations, that makes about 45,000 rules in total to be ported. This fits into the 64K words addressable by 16 bits. The 3D version however, with its (estimated) 150,000 rules, and its 24 symmetry rotations, will require over 3 million rules in total. The 3D version may require a "Super CAM" to be designed and built (by NTT's "Evolutionary Technologies" Dept., with whom the author collaborates closely), which can handle a much larger number of bits than 16. The group at MIT who built CAM8 is thinking of building a CAM9 with 32 bits. This would be very interesting to the author. Whether NTT or MIT get there first, such a machine may be needed to put the 3D version into a CAM. However, with a state-of-the-art workstation (e.g. a DEC Alpha, which the user has on his desk) and a lot of memory (e.g. 256 Mbyte RAM), it will still be possible to perform some interesting evolutionary experiments in 3D CAM-Brain, but not with the speed of a CAM.

Another possibility for porting the 3D version to the CAM8, is to re implement it using CA rules which are more similar to those used in von Neumann's universal constructor/calculator, rather than Codd's [1]. Von Neummann's 2D trails are only 1 cell wide, whereas Codd's 2D version are 3 cells wide, with the central message trail being surrounded by two sheath cells. The trick to using von Neumann's approach is incorporating the direction of motion of the cell as part of the state. The author's colleague Jacqueline Signorinni advises that CAM-Brain could be implemented at a higher density (i.e. more filled CA cells in the CA space) and without the use of a lookup table. The control of the new states would be implemented far more simply she feels, by simple IF-THEN-ELSE type programming. "von Neumann-izing" the 3D version of CAM-Brain might be a good task for the author's next grad student.

With the benefit of hindsight, if the 3D version is reimplemeted (and it is quite likely that my boss will have other members of our group do just that), then the author would advise the following. If possible (if you are implementing a Codd version [1]) give the four sheath cells in a 3D CA trail cross section the same state. This would obviously simplify the combinatorial explosion of the number of collision cases during synapse formation. But, how then would the 3D growth instructions be interpreted when they hit the end of a trail, and how would you define the symmetry rotations? If possible, it would also be advisable to use the minimum number of gating cell states at growth junctions for all growth instructions. Whether this is possible or not, remains to be seen. However, if these simplifications can be implemented (and of course the author thought of them originally, but was unable to find solutions easily), then it is possible that the number of 3D CA rules might be small enough to be portable to the CAM8, which would allow 3D neural circuits to be evolved at 200 million CA cells per second (actually less because of the subcell phenomenon).

Once the 3D rules are ready, two immediate things need to be done. One is to ask ATR's graphics people to display these 3D neural circuits in an interesting, colorful way, perhaps with VR (virtual reality) 3D goggles with interactivity and zoom, so that viewers can explore regions of the dynamic circuits in all their 500 colors (states). This could be both fun and impressive. The second thing is of course to perform some experiments on the 3D version. As mentioned earlier, this will have to be done on a workstation, until a SuperCAM is built. Another possibility, as mentioned earlier is to redesign the 3D CA rules, to simplify them and reduce their number so that they can fit within the 64K 16 bit confines of the CAM8 machine.

As soon as the 2D rules have been fully ported to the CAM8, experiments can begin at speed. Admittedly the 2D version is topologically different from the 3D version (in the sense that collisions in 2D are easier than in 3D), it will be interesting to try to build up a rather large neural system with a large number of evolved modules (e.g. of the order of a hundred, to start with). At this stage, a host of new questions arise. Look at Figure 6.14, which is van Essen's famous diagram of the modular architecture of the monkey's visual and motor cortex, showing how the various geographical regions of the brain (which correspond to the rectangles in the figure, and to distinct signal processing functions) connect with each other. Physiological techniques now exist which enable neuro-anatomists to know which distinct cortical regions connect to others. Thus the geography (or statics) of the biological brain is increasingly known. What remains mysterious of course, is the dynamics. How does the brain function.

Van Essen's diagram is inspirational to the author. The author would like to produce something similar with CAM-Brain, i.e. by evolving neural modules (corresponding to the rectangles, or parts of the rectangles) and their interconnections. This raises other questions about sequencing and control. For example, does one evolve one module and freeze its circuits and then evolve another module, freeze its circuits and finally evolve the connections between them, or does one evolve the two modules together, or what? Will it be necessary to place walls around each module, except for hand crafted I/O trails? The author has no clear answers or experience yet in these matters. The author's philosophy is "first build the tool, and then play with it. The answers will come with using the tool".

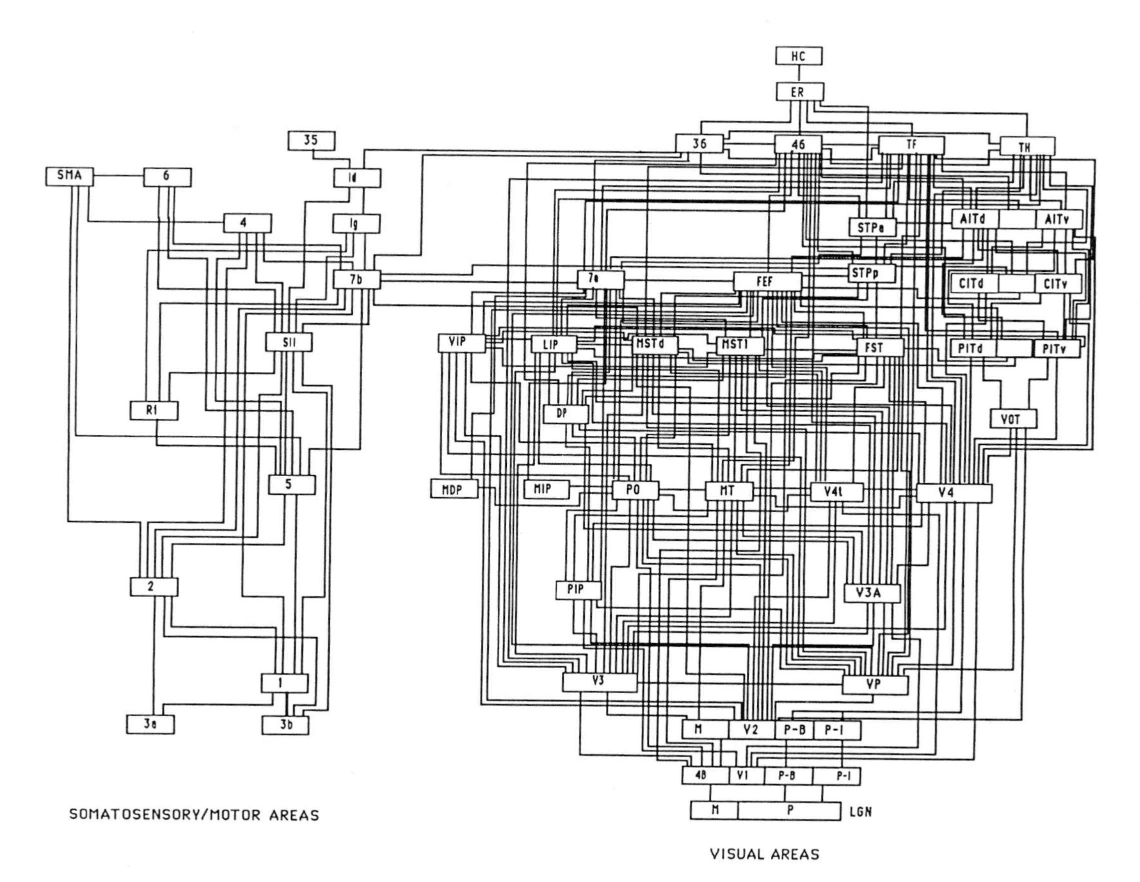

Figure 6.14: Van Essen's Monkey Brain Architecture

Another possibility for future work is to try to simplify the whole process of rule making. Perhaps higher level rules can be made which are far fewer in number and allow the author's low level rules to be generated from them. If such a thing can be done, it would be nice, but the author believes there are still so many special cases in the specification of 3D CAM-Brain, that the number of high level rules may still be substantial. If these high level rules can be found, it might be possible to use them and put them on the CAM8, so that 3D evolutionary experiments can be undertaken at CAM8 speeds. Another idea is to use FPGAs (field programmable gate arrays) which code these high level rules and then to use them to grow 3D neural circuits. Each 3D CA cell could contain pointers to its 3D neighbors. In this way, it would be possible to map 3D neural circuits onto 2D FPGAs. This is longer term work. FPGAs are not cheap if many are needed. The author's RAM based solution has the advantage of being cheap, allowing a billion (one byte) CA cell states to be stored reasonably cheaply.

A recent suggestion coming from NTT concerns the use of an existing "content addressable memory" machine, which may be able to update CA cells effectively. There is a "CAMemory" research group at NTT that ATR is now collaborating with. If a small enough number of CAMemory Boolean function rules corresponding to CAM-Brain can be found (a big if), it is possible that a NTT's CAMemory could be thousands of times faster than the CAM8. Obviously, such a possibility is worth investigating, and if successful, could be extremely exciting, since it would mean *hundreds of billions* of CA cell updates a second.

The author feels that the nature of his research in 1996 will change from one of doing mostly software simulation (i.e. generating masses of CA rules), to learning about the biological brain (i.e. reading about brain science to get ideas to put into CAM-Brain), hardware design, and evolvable hardware. These activities will proceed in parallel. Of course, evolutionary experiments, on CAM8 for the 2D version of CAM-Brain, and on a 256 Mbyte RAM (DEC Alpha) workstation for the 3D version, will also be undertaken in parallel.

Further down the road, will be the attempt to design a "nanoCAM" or "CAM2001" based on nanoelectronics. The Brain Builder Group at ATR is collaborating with an NTT researcher who wants to build nano-scale cellular automata machines. With the experience of designing and building a "SuperCAM", a nanoscale CAM should be buildable with several orders of magnitude greater performance. Further research aims are to use CAs to make Hebbian synapses capable of learning. One can also imagine the generation of artificial "embryos" inside a CA machine, by having CA rules which allow an embryological "unfolding" of cell groups, with differentiation, transportation, death, etc. resulting in a form of neuro-morphogenesis similar to the way in which biological brains are built. The author's "CAM-Bryo" program is an early example of this kind of neuro-morphogenetic research.

6.9 SUMMARY

The CAM-Brain Project at ATR, Kyoto, Japan, intends to build/grow/evolve a cellular automata based artificial brain of a billion artificial neurons at (nano-)electronic speeds inside Cellular Automata Machines (CAMs) by the year 2001. Quoting from a paper by Margolus and Toffoli of MIT's Information Mechanics group, "We estimate that, with integrated circuit technology, a machine consisting of a trillion cells and having an update cycle of 100 pico-second for the entire space will be technologically feasible within 10 years" (i.e. by 2000) [10]. In a trillion 3D CA cells (cubes), one can place billions of artificial neurons. Such an artificial nervous system will be too complex to be humanly designable, but it may be possible to evolve it, and incrementally, by adding neural modules to an already functional artificial nervous system. In the summer of 1994, a 2D simulation of CAM-Brain using over 11000 hand crafted CA state transition rules was completed, and initial tests showed the new system to be evolvable. By early 1996, a 3D simulation will be completed.

If the CAM-Brain Project is successful, it will revolutionize the field of neural networks and artificial life, because it will provide a powerful new tool to evolve artificial brains with billions of neurons, and at electronic speeds. The CAM-Brain Project will thus produce the first Darwin Machine, i.e. a machine which evolves its own architecture. The author is confident that in time a new specialty will be established, based partly on the ideas behind CAM-Brain. This specialty is called simply "Brain Building".

The author and his colleague Felix Gers are about to port the 2D version of CAM-Brain to the CAM8. Hence in early 1996, it will be possible to evolve neural circuits with 25,000 neurons (or 100,000 neurons, with 16 Mbit memory chips) at 200 million CA cell updates a second. As mentioned earlier, the author expects that when this happens, the world will sit up and take notice. Twenty years from now, the author envisages the brain builder industry (i.e. intelligent robots etc.) as being one of the world's top industries, comparable with oil, automobile, and construction. He sees an analogy between the efforts of the very early rocket pioneers (e.g. the American Goddard, and the German (V2) von Braun) and the US NASA mission to the moon which followed. Today's 100,000-neuron artificial brain is just the beginning of what is to come. With adiabatic (heat generationless) reversible quantum computation, it will be possible to build 3D hardware circuits that do not melt. Hence size becomes no obstacle, which means that one could use planetoid size asteroids to build huge 3D brain like computers containing ten to power 40 components with one bit per atom. Hence late into the 21st century, the author predicts that human beings will be confronted with the "artilect" (artificial intellect) with a brain vastly superior to the human brain with its pitiful trillion neurons. The issue of "species dominance" will dominate global politics late next century. The middle term prospects of brain building are exciting, but long term they are terrifying. The author has written an essay on this question [7]. If you would like to be sent a copy, just email him at degaris@hip.atr.co.jp (The author will set up his home page on the web in 1996, after making the effort to learn html).

Finally, by way of a postscript - as the author was preparing the final draft, there were 6 people at ATR working on CAM-Brain (the author (3D CA rules), and his colleague Felix Gers (porting 2D to CAM-8), the author's Japanese colleague Hemmi and his programmer assistant Yoshikawa (translating CA rules to Boolean expressions), and two M. Sc. students from Nara Institute of Science and Technology (NAIST). At NTT, there were 3-4 people from the Content Addressable Memory machine group who were finding ways to apply their machine to CAM-Brain. So, things are certainly hotting up.

(Note added, June 1996) - Figure 6.15 shows about 800 artificial neurons with their axons and dendrites grown using the CAM-8 machine with 128 Mega words of 16 bits. This figure is taken from an 8 square meter poster containing 100,000 neurons. In a year, this number will probably be a million. Felix Gers thinks he can port the 3D version to the CAM-8. The 3D rules are almost complete and number over 160,000, i.e. nearly 4 million with (24) rotations. Figure 6.16 shows a doubly zoomed 2D CAM-Brain on MIT's CAM-8. Figure 6.17 shows 2D CAM-Bryo and CAM-Brain on MIT's CAM-8.

By Xmas 1996, the author will have changed the architecture of CAM-Brain to a unit cell of 2 CA cells wide instead of six. The direction of motion of a CA cell will be contained in the state, so sheath and gating cells will no longer be necessary. This should simplify things a lot, and will obviously allow a much greater density of neurons. The number of rules for the 3D version will hopefully be less than 64K, so should run quickly on the CAM-8. A single neuron in this new architecture will be represented by a single CA cell. The author expects to put up to about *10 million* such neurons into a CAM-8 CA space of 128 Megawords of 16 bits each, by Xmas 1996. By anyone's standards, this should be more than enough to start building quite substantial artificial brains, and to

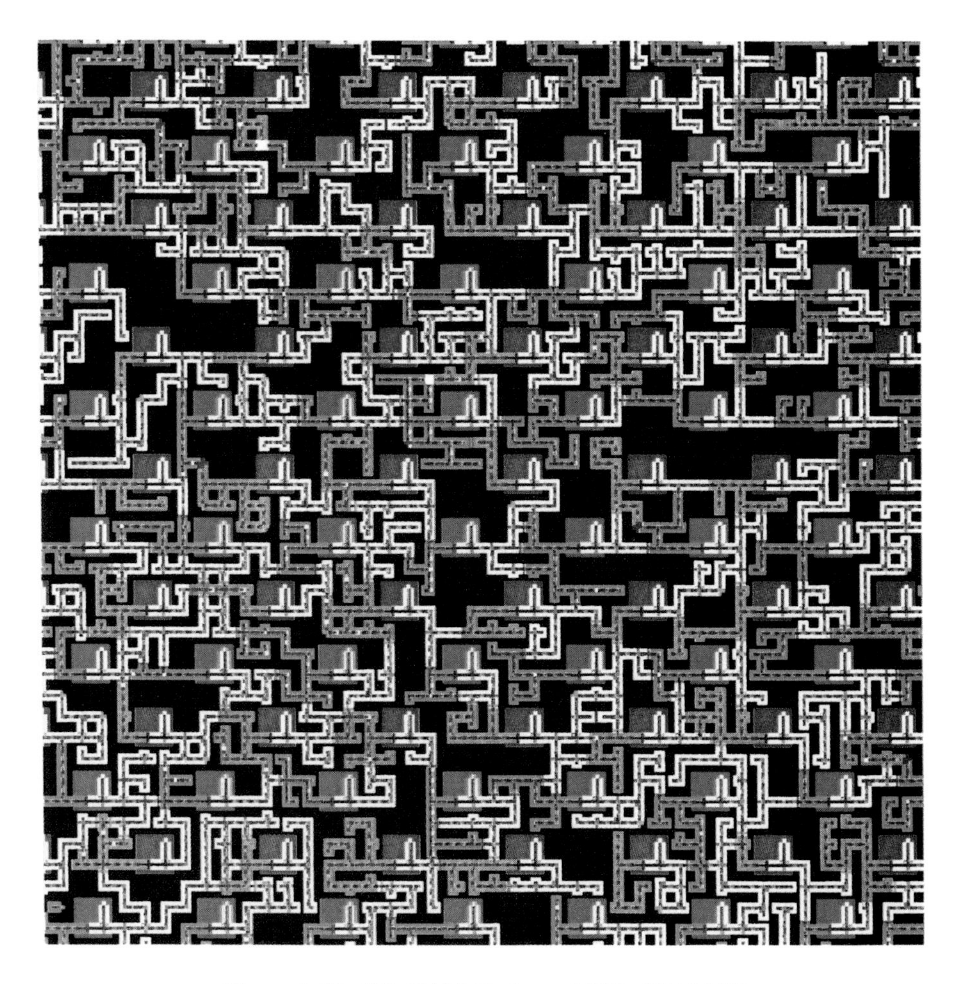

Figure 6.15: 2D CAM-Brain on MIT's "CAM-8"

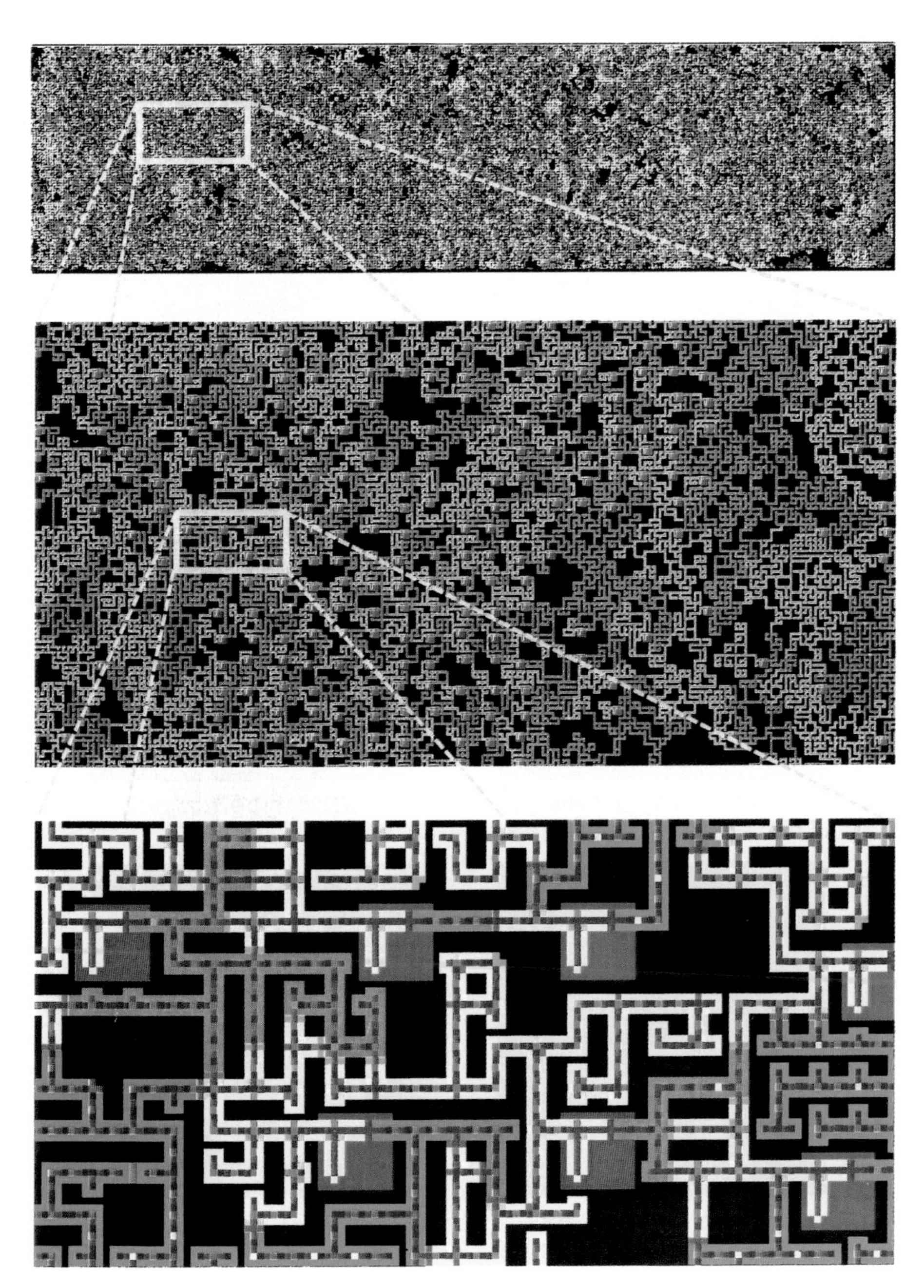

Figure 6.16: Doubly Zoomed 2D CAM-Brain on MIT's "CAM-8"

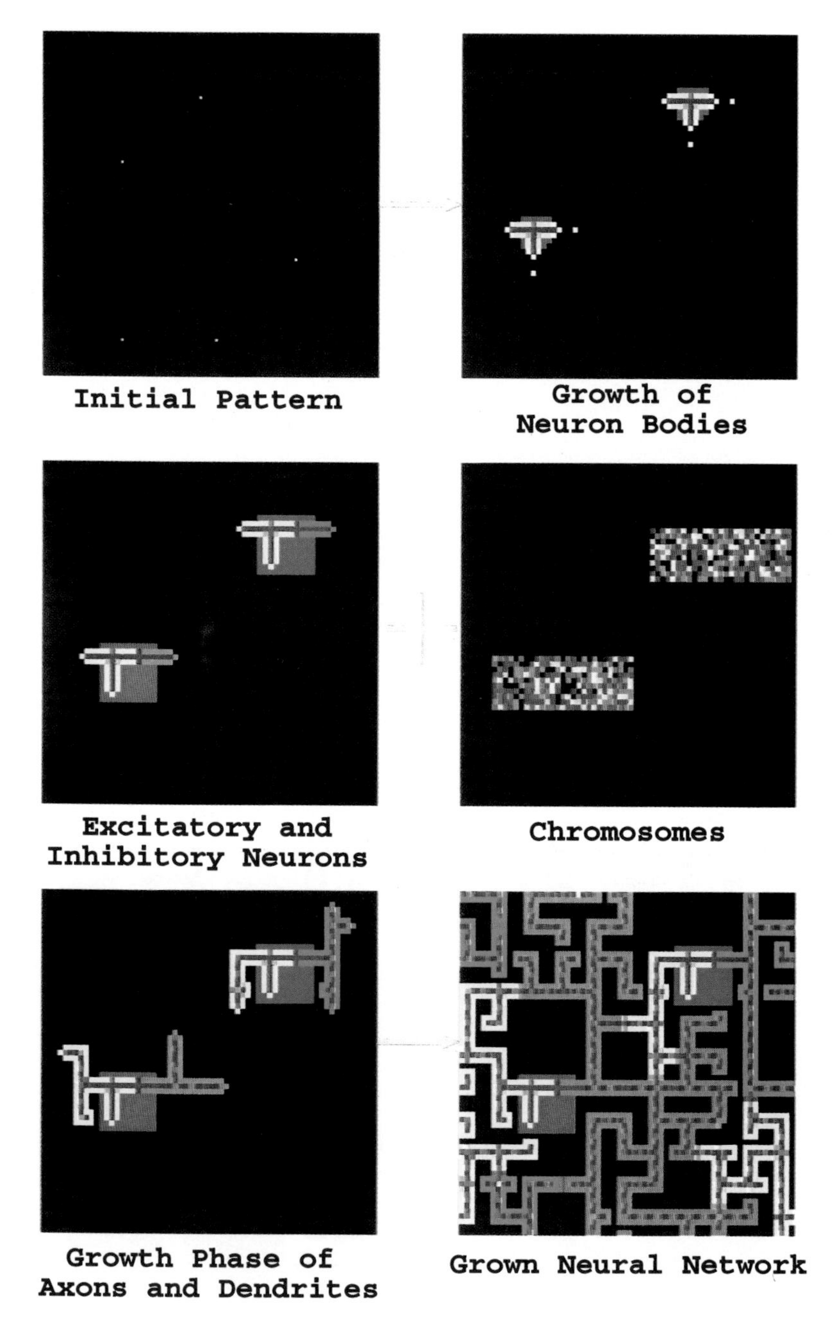

Figure 6.17: CAM-Bryo & CAM-Brain on MIT's "CAM-8"

revolutionize the field of neural nets, which is still stuck in its "single net syndrome".

Bibliography

[1] Codd, E.F., Cellular Automata, Academic Press, NY, (1968)

[2] De Garis, H., "Genetic Programming: Modular Evolution for Darwin Machines,"ICNN-90WASH-DC, (Int. Joint Conf. on Neural Networks), January 1990, Washington DC, USA.

[3] De Garis, H., "Genetic Programming", Ch.8 in book Neural and Intelligent Systems Integration, ed. Branko Soucek, Wiley, NY (1991)

[4] De Garis, H., "Artificial Embryology : The Genetic Programming of an Artificial Embryo", Ch.14 in book Dynamic, Genetic, and Chaotic Programming, ed. Branko Soucek & the IRIS Group, Wiley, NY (1992)

[5] De Garis, H., "Evolvable Hardware : Genetic Programming of a Darwin Machine", in Artificial Neural Nets and Genetic Algorithms, R.F. Albrecht, C.R. Reeves, N.C. Steele (eds.), Springer, NY (1993)

[6] De Garis, H., "Genetic Programming : Evolutionary Approaches to Multistrategy Learning", Ch.21 in book "Machine Learning : A Multistrategy Approach, Vol.4", R.S. Michalski & G. Tecuci (eds), Morgan Kauffman (1994)

[7] De Garis, H., "Cosmism : Nano Electronics and 21st Century Global Ideological Warfare", (to appear in a future nanotech book)

[8] Drexler, K.E., Nanosystems : Molecular Machinery, Manufacturing and Computation, Wiley, NY (1992)

[9] Goldberg, D.E., Genetic Algorithms in Search, Optimization, and Machine Learning, Addison-Wesley, Reading, MA (1989)

[10] Toffoli,T. & Margolus, N., Cellular Automata Machines, MIT Press, Cambridge, MA, 1987; and Cellular Automata Machines, in Lattice Gas Methods for Partial Differential Equations, SFISISOC, eds. Doolen et al, Addison-Wesley (1990)

Chapter 7

Fusion technology of neuro, fuzzy, GA and chaos theory and applications

Ryu KATAYAMA & Kaihei KUWATA
Sanyo Electric Co. Ltd., Hypermedia Research Center
Human Interface & Architecture Department
1-18-13, Hashiridani, Hirakata, Osaka, 573, Japan
{katayama,kuwata}@luna.hr.hm.rd.sanyo.co.jp

Lakhmi C. JAIN
Knowledge-Based Intelligent Engineering Systems
University of South Australia
Adelaide, The Levels, SA, 5095, Australia
etlcj@levels.unisa.edu.au

Recent advances in the theory and applications of artificial neural networks also called neural networks, fuzzy systems, genetic algorithms (Gas) and chaos theory have generated tremendous interest in fusion technology. This chapter presents the fusion technology of neural network and chaos, fuzzy system and chaos, genetic algorithms and chaos and the applications of chaos theory.

7.1 INTRODUCTION

Like fuzzy logic, neural networks and Genetic Algorithms (GAs), a novel paradigm called chaos theory [2], [3], is now emerging as one of the key technology of soft computing [1]. The term "chaos" means, in the scientific field, the complicated and noise-like phenomena originated from nonlinearity involved in deterministic dynamical systems [4]. There is a growing interest to discover the law of nature hidden in these complicated phenomena, and the attempt to use chaos theory for engineering purpose is gaining momentum [4]-[11], [57], [62]. Chaos is not only an important research field in science, but also has potential to be applied in many fields such as information processing based on dynamical system theory [12], [13], control and suppression of chaotic vibrations in mechanical systems and electrical systems [14], [16], synchronisation of chaos and communication using chaos [15], nonlinear signal processing such as adaptive equalisation [17], plant control [48], analysis and estimation of biological phenomena [18], or even the prediction of economic market price and analysis of sociological phenomena [19], [56], [62]. This new interdisciplinary field is called "chaos engineering" or "engineering chaos" [5]-[9].

A new interdisciplinary field using the combination of chaos theory with neural network, fuzzy theory, and GA is rapidly growing [7]-[11]. For example, by analysing chaotic signals observed from natural and artificial complex systems and finding hidden deterministic rules by use of fuzzy rules and neuro fuzzy techniques, one can interpret and identify the deterministic laws using the linguistic rules, whose mathematical model is not easy to obtain. These approaches make it possible to predict, control, and identify complex phenomena which were regarded as noise in the past.

In this chapter, we survey fusion techniques of neural network, fuzzy theory and GA with chaos theory and several applications which are widely reported in the literature (refer fig. 7.1).

7.2 FUSION TECHNOLOGY OF NEURAL NETWORK AND CHAOS

Chaos has a close relation with oscillation phenomena and is observed in fluctuation of the heart beating [72], capillary vessels [73], electroencephalographic (EEG) of olfactory bulb of rabbits [75], EEG of brain wave [74] and so on [76]. The role of chaos in these phenomena is considered useful in reinforcement of learning, environmental adaptation by homeochaos, pattern/symbol translation and time coding of information.

Especially, EEG research on the olfactory bulb of rabbits by Freeman et al. suggests the role of chaos in pattern recognition, pattern/symbol translation and process of learning [75]. When rabbits learn the new odor, the collective voltage level of olfactory bulb indicates the existence of small chaos close to limit cycle. Learning another odor, the attractor with different phase is generated. When unknown odor is presented, memory is searched according to the itinerary orbit

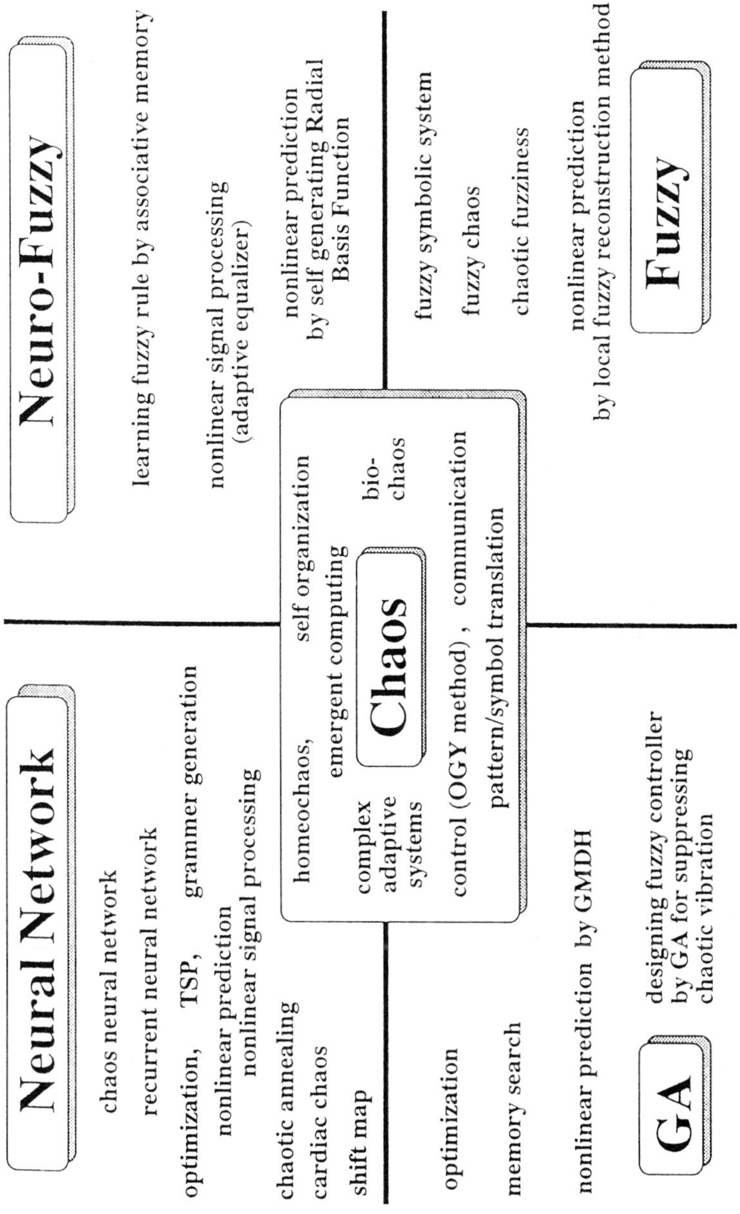

Neural Network

chaos neural network
recurrent neural network
optimization, TSP, grammer generation
nonlinear prediction
nonlinear signal processing
chaotic annealing
cardiac chaos
shift map

Neuro-Fuzzy

learning fuzzy rule by associative memory
nonlinear signal processing
(adaptive equalizer)
nonlinear prediction
by self generating Radial
Basis Function

Chaos

homeochaos, self organization
emergent computing
complex
adaptive
systems
bio-
chaos
control (OGY method), communication
pattern/symbol translation

optimization
memory search
nonlinear prediction by GMDH

GA

designing fuzzy controller
by GA for suppressing
chaotic vibration

fuzzy symbolic system
fuzzy chaos
chaotic fuzziness
nonlinear prediction
by local fuzzy reconstruction method

Fuzzy

Fig. 7.1 Fusion technology of neural network, fuzzy, GA and chaos

around the attractors that represent the learned odor. If the learning is succeeded, then another new attractor is generated [12,71,75].

To explain the information processing in the brain, various expressions and algorithms at every level are required. On this point of view, the attractors mentioned above are considered to be as one of the inner representations of symbol and concept in the brain. Then, there emerges important and challenging subjects as follows [77].

(1) What are the kinds and numbers of attractors ?

(2) What is the shape of domain of attractors ?

(3) How particular concept relates to some attractor ?

After all, if these problems are solved, it may be possible to design a new type of information processing device whose computational principle is totally different from that of today's digital computer.

Now, let us describe several applications of chaos. In the field of bio-chaos, the health care system called CAP [73] developed by Tahara and Tsuda triggered research on applications of chaos. CAP can estimate the condition of one's health based on the shape of attractor of pulsation of capillary vessel of one's finger. They reported their findings as :

(1) In case of unstable mental and physical condition or physical disease, the chaos becomes simple and non-structural.

(2) In case of good health, the structure of chaos becomes complex and dynamic.

Moreover, there appear complicated roll type, distorted and screw-type attractors in the state space.

In the field of engineering chaos, several paradigms such as chaotic neural networks are intensively studied [78-81]. They are applied to the combinatorial optimisation problems such as TSP (Travelling Salesperson Problem). Nozawa showed that chaotic neural networks are effective for the global optimisation problems [79]. Chen applied transient chaotic neural network to TSP [80]. In this model, chaotic dynamics is temporarily generated for global search. This chaotic dynamics gradually converges to the gradient dynamics with Lyapunov stability.

This process is called chaotic annealing and applied to the maintenance scheduling problem in an electric power system.

7.3 FUSION TECHNOLOGY OF FUZZY AND CHAOS

Research on the fusion technology of fuzzy and chaos theory was initiated by Diamond [23]. In order to extend state space from real vector space to fuzzy set space and consider dynamical systems defined on the fuzzy set space, it is necessary to introduce some kind of topology in the fuzzy set state space. Diamond introduced Haudorff metric and established sufficient conditions for mapping from real vector space to fuzzy set space which is chaotic, and called it "fuzzy chaos" [23], [26]. The conventional chaos or chaotic mapping defined is also called "crisp chaos" [27]. Kloeden extended Diamond's work and derived sufficient conditions defined by difference equation from fuzzy set space to chaotic fuzzy set space [24], [25].

In their experimental research, Horiuchi et al. considered a sequence which is obtained by iterative mapping of fuzzy sets, and classified several operators such as t-norm, averaging, and absolute difference by examining the dynamical behaviour of the sequence [28]. Yamakawa et al. analysed the transition of fuzzy information and shape of membership functions in the iterative fuzzy inference system with extension principle [29]. Grim regarded a multiple fuzzy inference system with self reference as a difference dynamical system, and examined the behaviour of fuzzy information using liar's paradox [30]. He also demonstrated that strange attractors and fractal structures are observed in these iterative systems.

In conjunction with fuzzy sets and fractals, Ezawa et al. introduced fuzzy Koch curve and fuzzy Julia Set, where iterative mapping constants are given by triangular fuzzy numbers [31]. They also showed that the fractal dimensions for fuzzy Koch curves are given by fuzzy numbers.

A number of researchers in the area of fuzzy and chaos have also restricted the chaos system parameters in their analysis. For example, Teodorescu treated "chaotic fuzzy systems", where only characteristic parameters of membership functions are chaotic [27]. He also classified these membership functions into "strong chaotic"," and "weak chaotic" according to the complexity of their behaviour. Possible application areas are modeling of chaos in economy and sociological phenomena [19], [34], [56], [62].

Diamond's approach is a purely mathematical one where fuzzy sets are treated in points in Banach space. However, the physical meaning of fuzzy sets as state space vectors in dynamical systems, is not understood yet. On the other hand, it is worth while to consider a system where the meanings of symbols are given by fuzzy sets, and their meanings are dynamically changed based on the time evolution of dynamical systems [12], [13].

On this point of view, Sato et al. proposed a chaotic fuzzy associative memory system, where a chaotic memory search is applied in a fuzzy associative memory system to support human creative thinking [32].

Katai et al. discussed chaotic structures of fuzzy symbolic dynamics, in conjunction with the hierarchical structure of natural systems and problem solving by self organisation [33].

It is predicted that these approaches may initiate a new field such as symbol processing and knowledge information processing based on dynamical systems.

The OGY (Ott, Grebogi, York) method is a popular control method of chaos, see Ott et al. [14]. For the synchronisation of chaos and communication using chaos, refer to Pecola [15].

Since fuzzy control is in general a kind of nonlinear control, stability theory for nonlinear control systems [35] and stability theory for fuzzy control systems [36] are useful design guides to avoid chaos in fuzzy control systems. There is another study using Lyapunov spectrum for evaluating the extent of stability in inverted pendulum controlled by fuzzy logic [37]. TDOF (Two Degrees Of Freedom) control is also proposed to control a system where disturbance signal or plant dynamics are chaotic [16]. TDOF fuzzy control is also proposed [63], however, it is not applied to the control of chaotic systems, and thus it is considered to be an interesting future research topic. All of these techniques are concerned with design principles to avoid chaos in control systems

Other studies on fuzzy control systems and chaos are in the prediction of unknown disturbance and load combined with fuzzy control system [8], [48], and application of fuzzy logic system to control a bouncing ball, whose behaviour is chaotic [38].

7.4 FUSION TECHNOLOGY OF GENETIC ALGORITHM AND CHAOS

GA (Genetic Algorithm) is basically regarded as an optimisation method or a search method. One of the advantage of GA is that it does not require the gradient information of the objective function with respect to the decision parameters. Therefore, GA is a very robust technique for the design problems such as optimisation problem with nondifferentiable objective function and structure decision problem [82]. With this advantage, GA is applied to the nonlinear prediction of chaos and chaotic control problems. Iba [83] identified the GMDH(Group Method of Data Handling) polynomial for Mackey-Glass chaotic time series which is known as a model of a heart beating, where GA is applied to the selection of explanatory variables and the estimation of coefficients of the GMDH model.

Karr [38] used fuzzy logic to control a bouncing ball with chaotic behaviour and applied GAs for the design of fuzzy controllers.

Nara et al. simulated a recollection process of image patterns which are stored in the Hopfield neural networks [70]. They proposed an efficient memory search method in a high dimensional state space with 2400 dimension, where GAs are applied to generate efficient searching orbits in the state space.

7.5 APPLICATIONS OF CHAOS THEORY

A number of contributions on chaos on chaos engineering [4] - [9], [11] , [62] and fuzzy chaos [10], [26] ,[27], [57] are reported in the literature. The following is a list of application areas of chaos engineering [5], [9].

(1) chaos computing

> Chaotic neural network [5], [9], computational universality and generalised shift map, optimisation such as travelling salesperson problem, artificial life, edge of chaos, antichaos and adaptation.

(2) deterministic prediction based on nonlinear dynamics

> Nonlinear prediction of chaotic time series [39]-[53].

(3) identification and modeling of chaos

Monitoring and diagnosis of complex plants such as electric power network or nuclear reactor [6], nonlinear signal processing such as voice synthesis, noise reduction and adaptive equaliser [17], and system identification [41].

(4) bio chaos

Analysis and diagnosis of EEG, MEG, cardiac chaos [54] and capillary chaos [18], sensitivity and mental stress engineering [18].

(5) chaotic memory

Memory dynamics of brain and neural network [12], [13], [32].

(6) chaotic coding and decoding

Image data compression, fractal image processing [8], chaotic cryptogram [62] and chaotic signal communication [15].

(7) chaotic pattern recognition

Taste sensor, fractal image recognition and feature extraction [6].

(8) chaotic art

Chaotic computer graphics, fractal computer graphics, chaotic music.

(9) utilisation of chaotic fluctuation and application to consumer electronics

Home appliances using chaos theory such as kerosene fan heater using chaotic fluctuation [8]-[11], dish washer [60], washing machine [64].

(10) devices for generating chaotic signals

Electrical and electronic circuits [61], [8], optical laser system [15], [55], artificial membrane [6].

(11) generation and control of chaotic signals in nonlinear engineering systems

Control of chaotic signals, suppression of chaotic vibrations in mechanical systems and electrical systems [14], path planning and wandering of autonomous robot using chaos [6], control of chaotic plant with two degree of freedom controller [16].

Here, we present some recent applications of chaos. Yamakawa and his research team [61] have fabricated a chaotic chip for experimentation using polysilicon CMOS process. This chip is suitable for implementing various systems based on chaos theory. A hard wired ciphering system based on chaos chip [69] was implemented for demonstrating the application of chaos in communication systems. The ciphering system is used to protect information. The ciphering (transmitting) system consists of a chaotic signal generator and the ciphering transmission system. The deciphering system consists of the chaotic signal generator and the receiving circuit. It may be noted that this system requires complete reproducibility of the chaotic sequence. Pecora and his co-workers [65] have also reported the use the synchronised chaos circuit for a private communication. It involves adding a chaotic signal to a message signal in a transmitter. The encoded signal is mixed with a drive signal for transmission. At the receiving end, the chaotic signal is subtracted from the encoded signal to yield the original information. No one will be able to extract information from the received signal unless one has a same synchronised chaotic signal generator for subtracting it from the received signal to recover the message.

Mori and Urano [68] have reported the application of chaotic time series for short-term power system load forecasting. Kang and Park [67] have used the chaotic dynamics in music auto composition. It is demonstrated that the chaos can be used to implement human life activity in the world of music.

Kawashima and Nagashima [66] have used a chaos neural network for implementing an original melody in the world of music.

Concerning the diagnosis and monitoring complex systems, Tsuda et al. developed a medical instrument to analyse chaotic pulse waves in capillary vessels of a finger [18]. They discussed possibility that shapes of reconstructed strange attractor reflect mental or physical conditions.

With regard to the applications to the home appliances, Sanyo Electric Company announced a kerosene fan heater (refer to fig. 7.2) in June 1992, which is the first consumer electronic product in the world using chaos theory [11], [9], [8]. Until 1992, it was considered to keep room temperature constant. On the contrary, studies by Wyon [58] and other researchers suggest that subjects actually prefer temperature swings about the optimal set point to the constant temperature. On the basis of this idea, appropriate temperature swing patterns are computed by a simple equation which produces a kind of 1/f like fluctuations generated by intermittent chaos [59]. Psychological experiments with the PMV (Predictive Mean Vote) values of about 30 subjects showed that sensation with appropriate chaotic fluctuations is more comfortable than with conventional control [11].

Another application to home appliances is a dishwasher with a 2-link nozzle by Matsushita Electric Industry [60]. A nozzle of usual dishwashers is composed of one link. However, to increase washing efficiency, they adopted 2-link mechanism which enables the nozzle to move with higher degree of freedom than conventional ones. They analysed the behaviour of 2-link nozzle, with Poincare sections, circle map, the largest Lyapunov exponent and the correlation dimension, and washing experiments with real tableware suggested that the behaviour of the 2-link nozzle is quasi-periodic or possibly chaotic with washing capability higher than conventional ones.

Fig. 7.2 Kerocene fan heater using chaotic fluctuations [11]

Goldstar Electric Company developed a chaos-controlled washing machine [64] in 1993. In this washing machine, the shape of the attractor, which is reconstructed from the time series of water flow, is used to judge whether clothes are in the state

Science	Engineering
Strong AI	Weak AI (IA:Intelligent Amplifier)
Fuzzy Theory	Fuzzy Control
Biological Neural Network	Artificial Neural Network
Scientific Chaos	Engineering Chaos
Adaptive Complex Systems	Artificial Life, GA
Scientific Soft Computing	Engineering Soft Computing
Clarification of the principle of Intelligence	Imitation and Utilization of Intelligence

Fig. 7.3 Two approaches toward Soft Computing

of tangling or untangling. If the state is judged as tangling, a pulsator pushed up the clothes to make them untangled.

7.6 CONCLUSION

We have reviewed the fusion of chaos with networks, fuzzy and technology. However, it seems that the extent of fusion of these techniques is superficial.

There seems to exist two approaches toward soft computing, namely engineering soft computing and scientific soft computing (refer Fig. 7.3). The soft computing research, whose purpose is to imitate and utilise the intelligence of human, is now rapidly expanding its application field. On the other hand, the aim of scientific soft computing is considered to clarify the principle of the intelligence, which realises the adaptation and learning ability in the dynamically changing environments. On this point of view, it is important to study further the mechanism to convert the pattern information associated with the physical phenomena of environments to the meaningful symbols and concepts for each individual.

Fuzzy theory is a top down paradigm where language and symbols are combined with numeric and pattern information. On the contrary, in biological systems, pattern information processing and symbol processing are combined using the dynamics of massively parallel neural networks. Chaotic information processing is now regarded as one of the promising computational model of this biological information processing [12], [13]. Chaos can create diversity from very simple deterministic rules, and this mechanism is considered useful to realise the structural change of autonomous systems, as a mechanism to alter the sub goals of creatures. More intensive studies on fusion of chaos and neural networks, fuzzy theory, and GA are expected such that subjective performance index or preferences, namely the objectives to work in the dynamically changing environments, are described by the combination of fuzzy theory and information processing principle based on nonlinear dynamical systems.

Bibliography

[1] Zadeh, L.A., Fuzzy Logic, Neural Networks and Soft Computing, Proc. of 5th IFSA Congress, plenary talk, Seoul (1993)

[2] Devaney, R.L., An Introduction to Chaotic Dynamical Systems, Second Edition, Addison-Wesley Publishing Company (1989)

[3] Thompson, J.M.T., and Stewart, H.B., Nonlinear Dynamics and Chaos - Geometrical Methods for Engineers and Scientists, John Wiley & Sons (1986)

[4] Horiuchi, K., Koga, T., ed., Special Issue on Engineering Chaos, Trans. of IEICE, vol.E73, no.6 (1990)

[5] Aihara, K., Expanding applications of deterministic chaos in engineering, Scientific American (the Japanese language version), vol.22, no.3, pp.26-33 (1992)

[6] Aihara, K., ed., Applied and Applicable Chaos, Science-sha, Tokyo (1994)

[7] Aihara, K., ed., Neuro, Fuzzy and Chaotic Computing: Towards a New Generation of Analog Computing, Ohm-sha, Tokyo (1993)

[8] Aihara, K., and Tokunaga, R., ed., Application Strategy of Chaos, Ohm-sya, Tokyo (1993)

[9] Aihara, K., and Katayama, R., Chaos Engineering In Japan, Communications of the ACM, vol.38, no.11, pp.103-107 (1995)

[10] Katayama, R., Chaos and Fuzzy Control Systems, Control and Information, vol.38, no.11, pp.619-624 (1994)

[11] Katayama, R., Kajitani, Y., Kuwata, K., and Nishida, Y., Developing Tools and Methods for Applications Incorporating Neuro, Fuzzy and Chaos Technology, Computers and Industrial Engineering, vol.24, no.4, pp.579-592 (1993)

[12] Skarda, C.A., and Freeman, W.J., How Brains Make Chaos in Order to Make Sense of the World, Behavioral and Brain Sciences, vol.10, no.1, pp.161-195 (1987)

[13] Tsuda, I., Dynamic Link of Memory - Chaotic Memory Map in Nonequilibrium Neural Networks, Neural Networks, vol.5, pp.313-326 (1992)

[14] Ott, E., Grebogi, C., and Yorke, J.A., Controlling Chaos, Phys. Rev. Lett., 64, pp.1196-1199 (1990)

[15] Pecora, L.M., Overview of chaos and communications research, Proc. of SPIE, vol.2038, Chaos in Communications, San Diego, California, pp.2-25 (1993)

[16] Kameda, T., Aihara, K., and Hori, Y., Application of a TDOF Controller to Chaotic Dynamical Systems, Trans. of IEE Japan, vol.113-C, no.1, pp.43-49 (1993)

[17] Kuwata, K., Watanabe, M., and Katayama, R., Adaptive Equalizer Using Self Generating Radial Basis Function, Proc. of 3rd International Conference on Fuzzy Logic, Neural Nets and Soft Computing, Iizuka, pp.493-496 (1994)

[18] Tsuda, I., Tahara, T., Iwanaga, H., Chaotic Pulsation in Human Capillary Vessels and its Dependence on Mental and Physical Conditions, International Journal of Bifurcation and Chaos, vol.2, no.2, pp.313-324 (1992)

[19] Rosser, J.B., From Catastrophe to Chaos, A General Theory of Economic Discontinuites, Kluwer Academic Publishers (1991)

[20] Terano, T., Asai, K., Sugeno, M., ed., Fuzzy Systems Theory and Its Applications, Academic Press, San Diego, California (1992)

[21] Katayama, R., Recent Trends on Fuzzy Hardware in Consumer Electronics Products, Journal of Japan Society for Fuzzy Theory and Systems, vol.6, no.3, pp.481-488 (1994)

[22] Takagi, H., Fusion technology of fuzzy theory and neural networks - Survey and future directions, Proc. of the International Conference on Fuzzy Logic and Neural Networks, pp.13-26, Iizuka (1990)

[23] Diamond, P., Fuzzy Chaos, Preprint, Mathematics Department, University of Queensland (1977)

[24] Kloeden, P.E., Chaotic Mappings on Fuzzy Sets, Proc. of 2nd IFSA Congress, pp.20-25, Tokyo (1987)

[25] Kloeden, P.E., Chaotic iterations of fuzzy sets, Fuzzy Sets and Systems, 42, pp.37-42 (1991)

[26] Diamond, P., Chaos and Fuzzy Representations of Dynamical Systems, Proc. of 2nd International Conference on Fuzzy Logic & Neural Networks, pp.51-58, Iizuka (1992)

[27] Teodorescu, H.N., Chaos in Fuzzy Systems and Signals, Proc. of 2nd International Conference on Fuzzy Logic & Neural Networks, pp.21-50, Iizuka (1992).

[28] Horiuchi, K., and Takeoka, M., On Repeated Fuzzy Operators, Proc. of 7th Fuzzy System Symposium, pp.529-532, Nagoya (1991)

[29] Yamakawa, T., Uchino, E., Miki, T., and Nakamura, T., Transitions of Fuzzy Number Through Nonlinear Dynamical Systems, Proc. of 2nd International Conference on Fuzzy Logic & Neural Networks, pp.145-148, Iizuka (1992)

[30] Grim, P., Self-Reference and Chaos in Fuzzy Logic, IEEE Trans. on Fuzzy Systems, vol.1, no.4, pp.237-253 (1993)

[31] Ezawa, Y., Haji, K., and Uemura, T., Graphical Display of Fuzzy Fractals, Proc. of 10th Fuzzy System Symposium, pp.399-400, Osaka (1994)

[32] Sato, T., Ushida, H., Yamaguchi, T., Imura, A., and Takagi, T., Chaotic Memory Search in Fuzzy Associative Interface, Proc. of 3rd International Conference on Fuzzy Logic, Neural Nets and Soft Computing, pp.203-206, Iizuka (1994)

[33] Katai, O., Horiuchi, T., Sawaragi, T., Iwai, S., and Hiraoka, T., Chaos Structures of Fuzzy Symbolic Dynamics and Their Relation to Constraint-oriented Problem Solving, Proc. of 10th Fuzzy System Symposium, pp.395-398, Osaka (1994)

[34] Teodorescu, H.N., Chaotic Fuzzy Models in Economy, Proc. of 2nd International Conference on Fuzzy Logic & Neural Networks, pp.153-155, Iizuka (1992)

[35] Narendra, K.S., and Annaswamy, A.M., Stable Adaptive Systems, Prentice Hall, New Jersey (1989)

[36] Tanaka, K., and Sugeno, M., Stability Analysis and Design of Fuzzy Control Systems, Fuzzy Sets and Systems, vol.45, no.2, pp.135-156 (1992)

[37] Yamaguchi, H., A Stability of Fuzzy Control System near Unstable Point, Proc. of 8th Fuzzy Systems Symposium, pp.505-508, Hiroshima (1992)

[38] Karr, C.L., and Gentry, E.L., Control of a Chaotic System Using Fuzzy Logic, In Kandel, A., and Langholz, G., ed., Fuzzy Control Systems, pp.475-497, CRC Press (1994)

[39] Casdagli, M., Nonlinear Prediction of Chaotic Time Series, Physica D, 35, pp.335-356 (1989)

[40] Takens, F., Detecting Strange Attractors in Turbulence, Lecture Notes in Mathematics, pp.366-381 (1981)

[41] Chen, S., Billings, S.A., Cowan, C.F.N., and Grant, P.M., Practical identification of NARMAX models using radial basis functions, Int. J. Control, vol.52, no.6, pp.1327-1350 (1990)

[42] Grassberger, P., and Procaccia, I., Characterization of Strange Attractor, Phys. Rev. Lett., 50, pp.346-349 (1983)

[43] Sugihara, G., and May, R., Nonlinear Forcasting as a Way of Distinguishing Chaos from Measurement Error In Time Series, Nature, vol.344, no.19, pp.734-741 (1990)

[44] Farmer, J.D., and Sidorowich, J.J., Predicting Chaotic Time Series, Phys. Rev. Lett., 59, 8, pp.845-848 (1987)

[45] Jimenz, J., Moreno, J.A., and Ruggeri, G.J., Forecasting on Chaotic Time series: A Local Optimal Linear-reconstruction Method, Phys. Rev. A, vol.45, no.6, pp.3553-3558 (1992)

[46] Mees, A.I., Dynamical Systems and Tesselations: Detecting Determinism in Data, International Journal of Bifurcation and Chaos, vol.1, no.4, pp.777-794 (1991)

[47] Ikoma, T., and Hirota, K., Nonparametric Regressive Model by Using Membership Function, Proc. of 8th Fuzzy System Symposium, pp.281-284, Hiroshima (1992)

[48] Iokibe, T., Kanke, M., Fujimoto, Y., and Suzuki, S., Short-term Prediction on Chaotic Timeseries by Local Fuzzy Reconstruction Method, Proc. of 3rd International Conference on Fuzzy Logic, Neural Nets and Soft Computing, pp.491-492, Iizuka (1994)

[49] Lapedes, A., and Farber, R., Nonlinear Signal Processing Using Neural Networks, Technical Report, no.LA-UR-87-2662, Los Alamos National Laboratory (1987)

[50] Moody, J., and Darken, C.J., Fast Learning in Networks by Locally-Tuned Processing Units, Neural Computations, 1, pp.281-294 (1989)

[51] Chen, S., Cowan, C.F.N., and Grant, P.M., Orthogonal Least Squares Learning Algorithm for Radial Basis Function Networks, IEEE Trans. on Neural Networks, vol.2, no.2, pp.302-309 (1991)

[52] Ikoma, T., and Hirota, K., Non-linear Autoregressive Model by Using Fuzzy Relation based Approximate Function, Proc. of 8th Fuzzy System Symposium, pp.277-280, Hiroshima (1992)

[53] Katayama, R., Kajitani, Y., Kuwata, K., and Nishida, Y., Self Generating Radial Basis Function as Neuro-Fuzzy Model and its Application to Nonlinear Prediction of Chaotic Time series, Second IEEE International Conference on Fuzzy Systems, San Francisco, pp.407-414 (1993)

[54] Mackey, M.C., and Grass, L., Oscillation and Chaos in Physiological Control Systems, Science, vol.197, pp.287-289 (1977)

[55] Ikeda, K., Daido, H., Akimoto, O., Optical Turbulence: Chaotic Behavior of Transmitted Light from a Ring Cavity, Phys. Rev. Lett., vol.45, no.9, pp.709-712 (1980)

[56] Matsuba, I., Application of Neural Sequential Associator to Long-Term Stock Price Prediction, Proc. of IJCNN'91 Congress, Singapore, pp.1196-1201 (1991)

[57] Yamakawa, T., Teodorescu, H.N., Unicho, E., and Miki, T., A Perspective on Application of Chaotic Fuzzy Logic Systems, First European Congress on Fuzzy and Intelligent Technologies, Aachen, pp.987-991 (1993)

[58] Whon, D.P., The Role of the Environment in Buildings Today: Thermal Aspects (factors affecting the choice of a suitable room temperature), Build International, vol.6, pp.39-54 (1973)

[59] Procaccia, I., and Schuster, H., Functional renormalization group theory of universal 1/f noise in dynamical systems, Phys. Rev., A28, pp.1210-1212 (1983)

[60] Nomura, H., Naito, E., Wakami, N., Kondo, S., and Aihara, K., Analysis of chaotic behavior of a 2-link nozzle in a dishwasher, Proc. of Japan-U.S.A. Symposium on Flexible Automation, ISCIE/ASME, pp.231-234 (1994)

[61] Yamakawa, T., Miki, T., Uchino, E., A Chaotic Chip for Analyzing Nonlinear Discrete Dynamical Network Systems, Proc. of 2nd International Conference on Fuzzy Logic & Neural Networks, pp.563-566, Iizuka (1992)

[62] Teodorescu, H.N., Engineering Applications of Chaos Theory, Tutorials of 3rd International Conference on Fuzzy Logic, Neural Nets and Soft Computing, pp.33-69, Iizuka (1994)

[63] Hayashi, S., Auto-tuning Fuzzy PI Controller, Proc. of 4th IFSA Congress, vol.Eng., pp.41-44, Brussels (1991)

[64] Kim, H.S., and Roh, Y.H., Implementation and development of chaos theory in washing machine, Journal of Korean Society of Mechanical Engineering, vol.34, no.6, pp.475-481 (1994)

[65] Ditto, W.L., and L.M. Pecora, Mastering chaos, Scientific American, August, pp.62-68 (1993)

[66] Kawashima, J., Nagashima, T., An experiment on arranging music by a chaos neural network, Proc. of the 3rd international conference on fuzzy logic, neural nets and soft computing, pp. 429-430, Iizuka (1994)

[67] Kang, H., Park, M.., Music auto composition using chaotic dynamics, Proc. of the international conference on fuzzy logic, neural nets and soft computing, Iizuka, pp.431-434 (1994)

[68] Mori, H., Urano, S., Chaotic time series analysis for short-term power system load forecasting, Proc. of the 3rd international conference on fuzzy logic, neural nets and soft computing, pp.441-442, Iizuka (1994)

[69] Teodorescu, N., et.al, Ciphering system using chaos devices, Proc. of the 3rd international conference on fuzzy logic, neural nets and soft computing, Iizuka, pp.655-656 (1994)

[70] Nara, S., Davis, P., Banzhaf, W., Functional performance of complex dynamics in neural networks, in Aihara, K., ed., Chaos in neural networks, Tokyo Denki University Press, pp.91-124 (1994)

[71] Tsuda, I., Chaotic brain, Science-sya (1990) (in Japanese)

[72] Goldberger, A.L., Regney, D.R., West, B.J., Chaos and fractals in human physiology, Scientific American, vol.262, no.2, pp.42-49 (1990)

[73] Tsuda, I., Tahara, T., Iwanaga, H., Chaotic pulsation in human capillary vessels and its dependence on mental and physical condition, International journal of bifurcation and chaos, vol.2, no.2, pp.313-324 (1992)

[74] Ikeguchi, T., Aihara, K., Chaos observed in electroencephalographic (EEG) of brain wave, in Aihara, K., ed., Chaos in neural networks, Tokyo Denki University Press, pp.91-124 (1994)

[75] Tsuda, I., Searching brain by chaos, Scientific American (the Japanese language version), vol.24, no.5, pp.42-51 (1994)

[76] Kaneko, K., Chaos giving diversity, Scientific American (the Japanese language version), vol.24, no.5, pp.34-41 (1994)

[77] Saito, T., Neural networks and chaos, Journal of Information Processing Society of Japan, vol.35, no.5, pp.42-51 (1994) (in Japanese)

[78] Aihara, K., Takabe, T., Toyoda, M., Chaotic neural networks, Physics letters A, vol.144, no.6,7, pp.333-340 (1990)

[79] Nozawa, H., A neural network model as a globally coupled map and application based on chaos, Chaos, vol.2, pp.377-386 (1992)

[80] Chen, L., Aihara, K., Chaotic Simulated Annealing and Its Application to a Maintenance Scheduling Problem in a Power System, International Symposium on Nonlinear Theory and Its Applications (NOLTA '93), Hawaii, USA, vol.2 (1993)

[81] Tani, J., Proposal of chaotic steepest descent method for neural networks and analysis of their dynamics, Trans. of the IEICE (Institute of Electronics, Information and Communication Engineers), vol.J74-A, no.8, pp.1208-1215 (1991) (in Japanese)

[82] Kobayashi, S., Present condition and problem of genetic algorithm, Journal of the Society of Instrument and Control Engineers, vol.32, no.1, pp.2-9 (1993)

[83] Iba, H., Genetic algorithm and information processing, Mathematical Science, no.363, pp.32-36, Science-sya (1993) (in Japanese)